D0126842

TENDING ADAM'S GARDEN

TENDING ADAM'S GARDEN

Evolving the Cognitive Immune Self

Irun R. Cohen

ACADEMIC PRESS

A Harcourt Science and Technology Company

San Diego   San Francisco   New York   Boston
London   Sydney   Tokyo

Copyright © 2000 by ACADEMIC PRESS

Academic Press
24–28 Oval Road, London NW1 7DX, UK
http://www.hbuk.co.uk/ap/

Academic Press
a division of Harcourt Brace & Company
525 B Street, Suite 1900, San Diego, California 92101-4495, USA
http://www.apnet.com

ISBN 0-12-178355-3

Library of Congress Catalog Card Number: 99-63116

A catalogue for this book is available from the British Library

Designed and typeset by Kenneth Burnley, Wirral, Cheshire
Printed in Great Britain by MPG Books Ltd, Bodmin, Cornwall

99 00 01 02 03 04 MP 9 8 7 6 5 4 3 2 1

To Yael:
lover, friend, partner, wife

About the Author

Irun R. Cohen was born in Chicago in 1937, and is married with three children and three grandchildren.

Professor Cohen studied Philosophy at Northwestern University, where he gained his BA in 1959, and subsequently studied Medicine at Northwestern University Medical School, gaining his MD in 1963. He trained in pediatrics at the Johns Hopkins Hospital, in Baltimore, and subsequently undertook research in infectious diseases as a commissioned officer of the US Public Health Service at the CDC, Atlanta.

He has been involved in research in the field of immunology at The Weizmann Institute of Science in Israel since 1968, where he holds positions as The Mauerberger Professor of Immunology, The Director of The Robert Koch–Minerva Center for Research in Autoimmune Diseases, The Weizmann Institute of Science, Rehovot, Israel, and The Director of The Center for the Study of Emerging Diseases, Jerusalem, Israel.

From the Pre-publication Quotes

'Thank you for sending me Professor Cohen's manuscript. I enjoyed it thoroughly. He has created a brilliant and iconoclastic text, that well-reflects his broad intellectual foundations. His novel ideas will be provocative for scholars and lay persons alike . . . (a) terrific work.'

Martin J. Blaser
Division of Infectious Diseases, Vanderbilt University, Nashville, Tennessee

'The book is beautiful as I suspected it would be, full of deep thought and philosophy. We are in need of your visions.'

Alex Whelan
St James's Hospital, Dublin, Eire

'(A) superb book . . . It ranks among the best dissertations on natural philosophy that I have ever read. Aristotle would be bemused and Darwin delighted.

'In a remarkably lucid presentation of the human immune system, Irun Cohen has rationalized it as a cognitive process; one capable of making decisions that preserve individuality and promote survival. Dr Cohen is not only one of the leading theorists of modern immunology, he is also an original natural philosopher – a meta-immunologist, meta-geneticist and meta-physician.

'Whereas modern molecular immunogeneticists have been discovering the nuts and bolts of the immune system, Dr Cohen has been absorbed as well in divining how the immune system has evolved, and its central role in evolution. He shows clearly how individuality is maintained by the receptors on immune cells. Through a genetically-evolved molecular self-recognition system these receptors create an internal picture of one's proteins that he dubs the immunologic "humunculus". At the same time, somatically modified (acquired) receptors recognize and protect against invading foreign molecules. Thus, a combined self and non-self recognition system has emerged.

'Self-recognition results in what Dr Cohen has popularized as "physiological autoimmunity"; autoimmune disease, on the other hand, is the pathological autoimmunity that occurs in a relatively small percentage of genetically vulnerable individuals who seem to have lost the ability to tolerate a limited number of autoantigens. Although the reasons for this intolerance are still obscure, the result is several major autoimmune diseases (e.g. rheumatoid arthritis, diabetes, systemic lupus erythematosus and multiple sclerosis). The possible blockade of such self-intolerance by vaccines is the focus of current intense research.

'Although Dr Cohen views the immune system as cognitive, and perhaps a model for the brain's cognition, he denies a preordained (teleologic) purpose of such cognition. Instead, he proposes that the immune system has evolved to preserve and nurture the future of mankind.

'And the spiritual quality of his biologic philosophy is reflected in his title, *Tending Adam's Garden*.'

Gene H. Stollerman
Professor of Medicine and Public Health, Emeritus, Boston University

'I consider the whole book to be quite brilliant.

'The first pages discuss basic concepts of evolution, causation, reductionism, determinism, emergence, information, meaning, chaos, attractors, cognition, etc. . . . (what have I left out of the list!) . . . with extraordinary clarity and brevity, and would provide huge stimulus for the large section of the public that is thirsting for books of IDEAS, and for interpretation of the backgrounds of science in accessible language . . . Just look at . . . Stephen Hawking's book . . . or the many books on chaos theory.

'This book shows how to apply to BIOLOGY (including the brain) "fashionable" concepts such as information theory, chaos theory and the behaviour of complex systems. He integrates all this with molecular biology and evolution in remarkably few words, and I am convinced that (these pages) . . . should be perfectly comprehensible to any intelligent reader, and could be presented to such readers as a series of essays on the organization of life.

'The whole book, including of course the second half (On Immunity) should be compulsory reading for all immunologists. As Irun Cohen clearly (but elegantly and gently) points out towards the end, it has been theoretically obvious for several years that the dogma taught to most students of immunology is based on an untenable hypothesis, and is holding up the development of realistic approaches to the clinical manipulation of the immune system.

'He writes so beautifully and clearly, with such crystallized wording, that even when I feel that the idea is not new, the way of saying it is a revelation.'

Graham A. W. Rook
University College London Medical School, London

'As one might have expected from such a sophisticated person, the book is very clearly written and elaborated, very attractive and full of original concepts on immunology and neurobiology. The text is deliberately simple, in a way which can be understood by a large spectrum of readers, including those with no or low scientific background. Much importance is given to psychological and psychosocial considerations.

'The topic is a fairly general one, involving major aspects of cognition approached at both levels of the brain and the immune system. A major effort is made to present the concepts in a very didactic fashion.

'The author is internationally known for his original ideas of self recognition by the immune system. The specificity of this book is to make a parallel between the author's ideas in immunology and their possible counterparts in neurobiology . . . This remains a hot debate . . .

'Publication of this book by Academic Press will honour the Company. The author's international reputation is outstanding and the text has been the matter of extensive thinking and care by an obviously gifted writer.'

Jean-Francois Bach
Necker Hospital, Paris, France

'It is TERRIFIC. Very clear, very well written, without a trace of cliché or second-hand thought. And amusing, full of good jokes and poetical statements. Most important, it seems to me to be true.'

N. Avrion Mitchison
University College London

'Thanks, Irun, it is wonderful to be asked to respond to writing so rich and capable of transporting me to realms I couldn't get to on my own.'

Eleanor Rubin
Museum of Fine Arts, Boston

'Cohen's book is a wonderful journey into the world of the immune system. The writing is original, provocative and touches on more than immunology, it touches on science itself.'

Howard L. Weiner
Harvard Medical School

Contents

Prologue

Prologue

PREPARATION

This book springs from two sources. One source is immunological, the other is personal. I started out intending to write a book describing my own special view of the immune system, a system I have been experimentally investigating for about 30 years. I wanted to write a book for a readership that would include not only immunologists and their associates, but also interested people with no specific background in the subject. Speaking clearly to a non-professional audience stimulates clear thinking, at least for me. Perhaps the recent emergence of grandchildren triggered a desire to clarify and summarize my professional thoughts. Whatever the motive, the book has turned out to deal with much more than plain immunology. On my way to launching the immunology, I found myself exploring more fundamental matters: causality, information, energy, evolution, cognition and individuality have entered the book, along with the more cloistered subjects of immunology.

Let me state why I think it is timely to write an immunology book aimed at a broad readership. The failings as well as the triumphs of the immune system have come to the attention of a public troubled by the need for new vaccines to control the spread of exotic viruses and drug-resistant bacteria, by the hope of immunizations to treat cancer, and by the increasing threat of autoimmune diseases like arthritis, diabetes, lupus and multiple sclerosis. The immune system functions to preserve health, and we want to know how to use it in our changing world. Immunology is no longer the private possession of a closed group of esoteric devotees, but a public commodity. Immunology has public obligations.

Immunology, too, needs pause for thought. Sciences, like people, are born and mature through differentiated stages of development. Immunology is now at a stage of taking stock. Modern immunology was born just over a century ago with the demonstration that the body contained cells capable of destroying the then newly discovered bacterial agents of infectious disease. It was proposed that evolution must have endowed the individual with a physiological mechanism – the immune system – responsible for actively defending the body against foreign invaders.

Since then, immunology has succeeded to a remarkable degree in fulfilling the quest of science for understanding basic mechanisms. Immunologists have penetrated the external cloak of immune events to discover the fundamental building-blocks that produce the events. For example, resistance to an infection can now be effectively reduced to a mechanism comprising a defined collection of cells, molecules and genes.

This process of reducing the immune system to its basic building-blocks was accompanied by a theory – the clonal selection theory of acquired immunity – that clearly pointed out the key molecular and genetic questions for experimental investigation. We will discuss the clonal selection idea in greater detail later in the book. For now, we need only note that the very success of the enterprise directed by the clonal selection theory has made the theory obsolete. However, in biology as in other sciences of the complex, obsolescence is a sign of success. The best experiments are those that produce news, the unexpected. So a productive theory ultimately will give birth to experiments whose unexpected results require a new theory to explain them. Thus a good scientific idea can be devoured by its offspring, which are the new facts generated by the idea. Indeed, a scientific theory that has survived unchanged into old age has not done its proper job.

The clonal selection theory has surely done its proper job. The litter of facts it has whelped can no longer be adequately nursed by the mother theory. The present need of immunology is not merely to discover yet more molecular building-blocks; at this stage we have an abundance, perhaps even a glut, of basic molecules. We now need to understand the way the molecules interact to create a system. Immunology is now sufficiently mature in facts to begin to consider organization, workings, behavior, and applications: in short, how the system fabricates complex behavior out of its simple building-blocks. Confounded by its very success, immunology finds itself in need of new unifying ideas, a new paradigm. The present developmental stage of immunology can be characterized as a transition from analysis (the reduction of observations to elemental building-blocks) to synthesis (the integration of the parts into a whole).

This book takes the synthetic path to develop the idea that the immune system is a cognitive system. The prototype of cognitive systems is the brain, and we shall discuss how the immune system, like the brain, learns through individual experience to organize both its internal structure and its external behavior. The cognitive paradigm suggests that we may be more successful in curing autoimmune diseases, in facilitating transplantation, or in designing effective vaccines once we more fully understand the decision-making processes of the immune system, and not merely its molecules.

THE READERSHIP

The book, as I said, is aimed at a broad group of readers. Professional immunologists and students of immunology should be stimulated by a new way of looking at familiar sights. People coming from other fields of biology might benefit from seeing the way the immune system does its business. Physicists with a leaning towards complex systems will be enlightened by considering how sophisticated behavior emerges from the interactions of simple molecules. Cognitive scientists and neuro-biologists can think about learning, decision-making, memory, and recognition done by another mode of operation. Philosophers of science and mathematicians will be offered paradigms. And any interested reader, curious about life, health and disease, can find some answers (and perhaps more questions) about the subject of immunity and its relationship to infectious diseases, vaccination, autoimmunity, transplantation, and cancer. In short, the book has been written to tell various readers not only what the immune system actually does, but how it does it. But how can one book talk sensibly to such a wide-ranging readership?

THE APPROACH

To illustrate its concepts and tell its story, the book draws on metaphors and examples from everyday life. I feel I can do this because all systems, be they social, biological or physical, employ common principles of organization. For this reason familiar and homely examples can be used to discuss seemingly complex scientific facts or principles. Metaphors and concrete examples can have clear meaning to people in diverse walks of life who share a common culture. Thus, the reader should not need a special expertise or a special vocabulary to understand the book.

Moreover, I have used a strategy of informing the reader through case histories and figurative metaphors, not only because I am an experimentalist and like to think concretely and associatively, but because I also want to mobilize the reader. When I invoke, for example, a river to illustrate concepts of emergence, history or evolution, the reader is invited to use the current of my thought to develop his or her own ideas about the subject. In fact, I invite the reader to row upstream and question my metaphors, producing better ones instead; without necessarily fathoming my mind, the reader may bring to the surface what's in the depths of his or her own mind. Metaphors, as the poets have shown us, can deepen thought by triggering associations between seemingly unrelated matters. Indeed, using mundane examples to clarify seemingly esoteric science can make the science tangible, and at the same time the association may stimulate the reader, as it has the author, to consider the scientific wonder of everyday experience.

THE PLAN

The book begins with a brief introduction to individuality and its expression in two systems: the brain and the immune system. This sets the stage for comparing both systems from a cognitive point of view. The body of the book unites five themes:

1. On Causality and Evolution.
2. On Cognition.
3. On Immunity.

4. On Autoimmunity.
5. On Tending Adam's Garden.

The first two themes, On Causality and Evolution, and On Cognition, draw the conceptual framework within which we shall position the immune system. On Immunity describes the molecular and cellular agents of the system and their organization in space and time. Our aim here is to tell how the system maintains and protects the individual. On Autoimmunity deals with the way the immune system relates to the individual body and defines its individuality. This theme also includes some basic principles related to the boundary between the self and the world expressed in immunity to cancer, the rejection of foreign tissue transplants, and the interactions between the self and infectious agents. The final section, On Tending Adam's Garden, considers some questions about the meaning of Adam's individuality and the husbandry needed to tend a changing Garden.

Each theme is composed of one or more subjects, and each subject is divided into a series of topics. The topics, described in one or a few paragraphs, are numbered consecutively (§1, §2, and so on) throughout the book. Certain concepts will be found to surface within the realms of different themes and subjects. Hence, within the exposition of a topic, related topics may be cited by number so that the reader can refer back (or ahead) to them at will. Indeed, a knowledgeable reader might begin the book at any point of interest and follow the thread of the argument by tracking citations to related topics. Thus the book may be explored as a network of connected ideas within which a reader might want to construct his or her own hypertext. Of course, the book has been written to be read, as other books, in continuity from cover to cover; but it need not be read that way.

I have not generally referenced the facts I present and do not cite the professional scientific literature. Readers who are professionals know the citations, and readers who are not professionals need not be burdened. Today, any interested person has easy access to the world of professional literature on the Internet. I beg the indulgence of any reader disappointed by this book's lack of traditional scholarship, irritated by its personal views, or offended by its failure to mention a name. I do refer, on occasion, to some of my experiments and to books, articles or people who have influenced my general thinking.

As the themes unfold, I will compare other ideas on the subject, classical or controversial, with those I have written. The contrast will allow the reader a balanced view from which he or she can raise questions and form opinions. The reader is free to view the prints of M. C. Escher, which decorate the book, and also illustrate some of its ideas.

To help the reader anticipate the drift of the topics within a subject, each subject is introduced by a short note, printed in *italics*.

HELPERS

The book was written with the aid of electronic mail consultations, dialogues and arguments with friends who include artists, professors of English, mathematicians, philosophers, physicists, evolutionary biologists, microbiologists, physicians and immunologists of various strains. Some of the most helpful friends have been my students. I wrote and re-wrote sections in response to their e-mailed comments and questions. I believe most of these friends finally were satisfied, convinced, or despaired of convincing me. Having learned friends, of course, does not relieve me of the responsibility for any faults in accuracy, depth, precision, or clarity. Special thanks are due to Mrs Doris Ohayon for helping me with the typing and to my colleague Johannes Herkel for help in preparing the pictures. Thanks too to Tessa Picknett, my executive editor, who listened to my story over a cup of coffee, and then made it happen. I am grateful to the Weizmann Institute of Science for cultivating a garden where ideas can flower. I hope the book will be a pleasure to read, as it has been to write.

Chapter 1
Introduction

Chapter 1
Introduction

ON INDIVIDUALITY

The story of immunity is the story of the self. The immune system, like the brain, creates, records and protects our individuality.

§1 Individual Brain

The world contains more than 5 billion people and no two of us are exactly the same. What is it that endows each of us with our individuality? What is it that makes each of us different from all the rest? The word 'individual' comes from the Latin *individuus* meaning 'that which cannot be divided'. The Latin-derived 'individual' has the same root meaning as the Greek-derived 'atom', the kernel of being that defines the thing in itself. The core of individuality is that which remains after all the extraneous trappings and baggage, all the incidentals and accidentals, have been peeled off the person. Individual means irreducible, nothing more can be stripped away; the last mask, the final costume have been removed from the person. Splitting an atom of matter changes its chemistry; despoiling the atom of individuality destroys the person. Individuality is the inner sanctum of the self. Of what stuff is this essential self made?

The core self is defined differently by prophets, philosophers, jurists, and psychologists, each in the light of a favored teaching. Artists explore the meaning of the self in word, tone, rhythm, movement, hue, line or surface. These different ways of looking at the essential self are products of the human central nervous system. Faith, thought, society and art are expressions of the human mind. Different views of individuality arise from the different ways human thought has looked at the products of human thought: mind, as it were, conscious of itself. Consciousness, then, is the internal perception of one's individuality.

The individuality of each human brain, moreover, can be seen by an external observer; individuality is not merely a figment of internal awareness. At the material level of biological structure, no two people have the same brains. Identical twins, who arise from the same fertilized egg, develop physically different brains. True, brains begin life expressing DNA information inherited from mother and father. DNA, nevertheless, is only the beginning. To function, brains have to progress beyond their DNA-encoded building-blocks. Brains actually modify their anatomy and physiology during the ongoing interaction of the individual with the environment. Brains do not merely *respond* to experience, they *require* experience. Brains deprived of input don't realize their potential. If corrected too late in life, children with a squint may become functionally blind, and children with a hearing defect, unless they are given special attention, will have difficulty in learning to speak properly. Thus the brain is built by experience. Seeing is required to form the faculty of sight, hearing is required to generate the capacity to hear meaning. This is the principle of self-organization, the principle that you get better at it by doing it. You build your brain when you use it; better, you build your brain only if you use it.

Use creates the capacity to use, because each brain builds a private set of synaptic connections and organizes individual networks of neurons (brain cells) in response to personal experience. The brain comprises about 10^{12} (a thousand times a billion) neurons, which have about 10^{15} connections (a thousand times more), so there is plenty of opportunity for fashioning individual variation. Since no two people, identical twins included, ever experience precisely the same neurological environment, the central nervous system of each of us is manifestly unique – functionally, chemically and anatomically. The brain is thus formed by its individuality.

Even if we dared clone people, as we have cloned sheep, the cloned people would develop into distinctly different individuals. Sports fans have dreamed of creating a basketball team consisting of five clones of Michael Jordan, a person reputed to be the greatest basketball player of all time. The fans, however, would be disappointed; the clones of Michael Jordan, each Michael Jordan developing his own brain, might not have either the talent or the interest to excel in the NBA (National Basketball Association). The individuality of the mind arrives with the biological individuality of the brain.

§2 Individual Immunity

But there exists another biological system that defines the individual: the immune system. The immune system has nothing to say about the spiritual, logical, legal or poetic self, but the immune system has much to say about the molecular self. By its acts, the immune system defines the material components that make up the self. The immune system is the guardian of our chemical individuality; it is a system that eliminates parasitic bacteria and viruses, a system that rejects foreign cells and tissues, a system that can destroy tumor cells arising from our own bodies. By deciding which macromolecules and cells are allowed residence within us, the immune system establishes the molecular borders of each person. In defending the individual, the immune system defines cellular individuality.

The immune system has earned a reputation, justly, for its role as protector of the body against foreign invaders. However, the immune system is not only a department of defense, it also functions as a department of internal welfare. The immune system is an unsung hero of maintenance and reconstruction. It deals, as we shall see, with humble bumps, shocks and grinds, and not only with catastrophic evils.

Like the central nervous system, the immune system is individualized. Identical twins, born with identical DNA, develop different immune systems, just as they develop different brains. Each person's immune system records a unique history of individual life, because, as we shall discuss further on, the immune system, like the brain, organizes itself through experience (§71–§74).

Thus the brain and the immune system establish individuality at two levels: they help us adapt to life and so preserve us, and they make a record of what has happened (§61). By recording each person's life experience, the two systems are custom-made, manufactured one of a kind for each. Like the brain, the immune system preserves the individual, and, in doing so, defines the individual. Our five cloned Michael Jordans would express their individual differences in their antibodies as well as in their thoughts.

§3 **Individual Concerns**

Immunity, however, is not only a blessing. One's immune system, like one's brain, can cause distress. The immune system, in addition to protecting us, can turn upon the body itself and damage vital organs in what are called autoimmune diseases. Multiple sclerosis, juvenile diabetes, rheumatoid arthritis and other maladies, which lately seem to be affecting more and more people, are caused by an immune attack directed against normal body components. Autoimmune diseases happen when our guardian becomes our antagonist (§146).

For better or worse, the molecular self depends on the behavior of the immune system, just as the spiritual, intellectual, emotional and social selves depend on the workings of the brain. To understand health and disease, we have to consider how the immune system treats its friends and its enemies, how it selects targets. Similar to our dealings with other service organizations, we pay no mind to immunity so long as it quietly does its job of protecting us. We notice the system most when it fails to serve, when we justly complain about newly emerging viruses such as HIV that evade our immune defenses, about needed transplants that get rejected, about tumors that don't get rejected, about auto-immune diseases that make life a misery, about allergies that annoy. We are incited to think by disease, not by ease. How the immune system behaves defines not only who we are, but how we feel. Our personal stake in the operation of the immune system is great and it is worthwhile to consider how it works to separate each of us from the rest of the world, how it makes us individuals. Quite simply, our lives depend on our immune individuality.

Chapter 2
On Causality and Evolution

Chapter 2
On Causality and Evolution

CAUSALITY

We shall begin with a discussion of causality because the concept of cause is fundamental to the world view of science. Guided by Aristotle's definition of causality, we shall see how science seeks to reduce the phenomena of the material world to underlying transformations of information and energy. The discussion of information and energy will set the stage for an understanding of the process of evolution. And we need to understand evolution to begin to understand the immune system.

§4 Aristotle's Causality

The concept of causality is important to any discussion of nature. We would like to know, for example, what has caused the immune system to evolve the way it has, and how immunity may cause health or disease. But indeed, we also want to cause the immune system to get rid of a virus or to refrain from rejecting a transplanted kidney. These are different uses of the concept of cause. What then is a cause?

Aristotle proposed that four factors converge to account for the existence of an entity: its matter, its form, its maker and its end, the goal for which the thing is meant. A cake, for example, exists by virtue of its ingredient matter (flour, sugar, eggs, water), its form as a cake rather than as a knot of spaghetti, the person who baked it, and the birthday party at which it is to be eaten.

Aristotle's four-tiered classification of causality is not the way science likes to think about causality today, but it is helpful to consult Aristotle because his classification calls attention to the key issues. Aristotle seems to be saying that knowledge of our cake requires not only that the cake before us exists (the material cause of the cake), but that the cake satisfies our idea of what a cake should be (the formal cause), which allows us to distinguish a good cake from a failed cake.

The third element in Aristotle's anatomy of causality, the efficient cause, is closest to the way we now think about causation. This element refers to the forces that actually bring the cake into existence: the baker who puts the ingredients together, the heat and time needed to bake the cake, and the cake-baking chemical reactions.

The fourth element in Aristotle's definition of causality is the trickiest and the most notorious, that of the final cause: the reason why the thing exists, what its function is, what it is good for. We call this factor teleology (from *telos*, *end* in Greek), the doctrine of final causation. People initiate actions leading to intended goals, but teleology imputes intentions to nature herself. We may be inclined to say that the proof of the cake is in the eating, but who can say what the function of this cake really is? Was it baked to celebrate a ritual (a birthday), to entice a child to take nourishment, to win a cake-baking contest, to earn a living, to . . . ? Ends are a matter of opinion, and so teleology is no longer considered a causal factor in science.

Aristotle's amalgamation of matter, agent, form, and end seems to lump together ontology (What is it?), epistemology (How do we know about it?) and teleology (What is it for?), along with the proximal cause (Who did it?). For Aristotle, the thing-in-itself is only the beginning of the story. Aristotle's classification of causality requires us to consider the role of the thing-in-the-world. Aristotle is a causal extrovert. Western science, in contrast, is introverted, it prefers to isolate its objects of investigation from the rest of the world and focus on their innards. What kinds of causal explanation suit our science?

§5 Reductionism

Minds thirst for simple causes to explain the complexities of the world. The cycle of the seasons perplexes: why do the days get shorter as we progress into winter, why does it get colder? The mind searches for a simplicity that accounts for the observed complexity of the change of seasons. To our satisfaction, the complex cycle of the seasons can be explained by the way the earth orbits the sun, and that circumnavigation, in turn, can be explained by the law of gravity. Metaphorically, we *reduce* the complexity of the seasons to causal principles that generate the complexity. (A lucid exposition of scientific reduction, simplicity, complexity and chaos can be had by reading *The Collapse of Chaos*, by Jack Cohen and Jan Stewart [1994].)

Reduction is central to the scientific method, but the search for underlying causes is not an invention of Western science. Reduction seems to be a habit of the human mind. Minds find satisfaction in reductions, chosen according to taste, that can explain the state of the world. Reductions can be mystical. The will of the gods, the juncture of the stars, retribution for sin, fate, luck, all have their adherents. By grasping a simple cause, we free ourselves of the annoyance of the complex. Reductionism is the principle of replacing perplexity (complex questions) with what passes for understanding (simple answers).

Assigning a cause tends to put minds at ease because causes have names. Naming is a handle, a type of owning, a type of artificial control. Names allow minds to compress a complicated, often threatening facet of reality into a manageable abstraction. Minds are quite adept at shuffling abstractions. Once we name it, we are free to treat it as we may. We find relief in just saying the word, in telling the story, in prayer. Names work magic.

The reduction of reality to verbal or nominal symbols is human, but reduction is scientific only when performed according to the empirical and logical methodology of science. The mental abstractions generated by scientific inquiry are not as simple as are most other reductions, for science demystifies; scientific reduction translates reality into the language of mathematics or into special causal relationships. What causal reductions are allowed?

§6 Reductions of Science

Scientifically reductionist causes can be classified into five sorts:

1. Fundamental laws of nature. The most basic of explanations are those that can be tied to the laws of nature uncovered by physics, such as gravity, electromagnetic and other elemental forces, the conservation of energy, atomic and molecular structure, relativistic mechanics, quantum mechanics, and so forth. These basic factors are timeless and their truth is independent of any particular instance or happening. Such laws and rules are fundamental because all particular instances and happenings obey them.

Living creatures are matter too and must behave according to the laws of matter. The fundamental laws of nature are necessary for life; however, such laws are not helpful when we wish to understand the forms or behaviors of living organisms. We do not answer biological questions by reduction to the fundamental laws of physics. For example, life is constrained by space/time, but particular manifestations of life cannot be meaningfully explained by the big bang that produced space/time. The big bang is necessary, but not sufficient to account for the appearance of life on earth. In the same vein, a coroner would lose his job were he to attribute the cause of death to the fact that the dead one had been alive. True, all multicellular creatures are destined to die. But this law of biology is too fundamental to be a satisfactory explanation for any particular instance or event. The mind does not like to accept explanations that are too grand or too far removed from the human scale. Causes, to be relevant, have to be the right size.

2. The agent–event relationship. The assignment of an agent as the cause of a biological event is a more useful scale of reductionism. The patient's acquired immune deficiency syndrome (AIDS) was caused by infection with the HIV virus, the broken hip was caused by a fall, the recovery from pneumonia was caused by penicillin, the divorce was caused by another woman, or man. Causality at the level of the agent, Aristotle's third factor in causality, can be quite satisfactory at the clinical, legal, or social levels. The coroner does his job well when he determines that the patient died of pneumonia, or that the subject was already dead of a heart attack at the time his car crashed. Agent–event causality, however, tends to leave questions of causality open at finer scales. How does HIV infection cause AIDS, how does penicillin kill gram positive bacteria, how does falling break bones weakened by osteoporosis? Some causal investigations, however, may come to a halt at the agent–event scale; the 'other' woman or man, for example.

3. The structure–function relationship. Most prized in biology research is the reduction of action to structure. Antibodies, enzymes and other biological molecules act the way they do thanks to the way they are constructed (§100–§102). Their activities are inherent in their shape and molecular make-up. Antibodies recognize their antigens and enzymes form or degrade their substrates because of their molecular character-istics. Morphine influences the brain, cortisone reduces inflammation, sleeping pills induce sleep, bacterial toxins kill, all because of their molecular structures. Understanding relationships between structure and function is the molecular basis for understanding physiology and disease; it is the key to designing drugs and other therapies. This scale of causality provides control. Note that the word 'function', in this context, denotes only the demonstrable activity of the structure, not its teleology. The word 'function' here does not imply purpose.

4. *Energy transactions.* Structure–function relationships do not resolve all issues. Some questions are better reduced to transfers of energy from cause to effect (§9). The billiard ball enters the pocket by the transfer of a momentum vector from the cue. The runner wins the race by the enzy-matic transfer of energy from glucose molecules to muscle contractions. Energy transfers, rapid and slow, can also be destructive: bullets, explo-sions, falls, fire, wind, rusting, decomposing, ageing. Energy transactions are fine-scale causes in biology, chemistry and physics. Indeed, energy transfers express fundamental laws of physics, but geared to a scale of discrete events.

5. *Information transfers.* Perhaps more fundamental than energy trans-fer in the process of causality is the transfer of information. Information has a formal definition, so let us define it.

§7 Information

The word 'information' is derived from the Latin *forma,* which means a mold or a representation. Information is related to Aristotle's con-cept of *form,* the characteristic structure of an entity that sets it apart from other things (§4). Information thus connotes a distinctive arrange-ment. The word 'information' is used freely, and most people who use the word feel no need to stop and define its meaning. However, the rapid development of communications technology beginning in the middle of the twentieth century drew attention to the fact that infor-mation was not merely the fabric of intelligence; it was also a commodity to be bought and sold. Information merited quantitative consideration.

Among those who sought to develop a scientific theory of information was Claude Shannon. Shannon, a scientist at the laboratories of the Bell Telephone Company, was concerned with the fidelity of systems transmitting messages. Shannon (and the telephone company) wished to design the optimum way, at least in theory, for effecting the faithful transmission of a message from a source to a receiver through some channel of communication. The aim was to make sure that the *output* of the communication channel, be it wired or wireless, was a reliable copy of the *input*, free of distortion and error. The task was to ensure that the information inserted at the input would emerge at the output. The message, irrespective of what it may mean to the sender or the receiver, has to be a just-so arrangement. But what is an arrangement? How can one assign a quantity to a formal arrangement, or compare degrees of arrangement? The answer to that question was Shannon's insight into information.

Shannon's idea was to compute the *amount* of information inherent in a message as the degree to which the message is *just-so*. The concept of *just-so* agrees nicely with the root meaning of information as a mold, and with Aristotle's idea of form as a distinctive characteristic (§4). According to Shannon, the degree to which an arrangement is just-so depends on the number of other possible ways the elements comprising the message might otherwise have been arranged. For example, if the message happens to be a string of letters (which may or may not create a meaningful sentence), then the alternative arrangements to that particular string may be defined as all the other possible combinations of the letters falling within the format of that message. Or if the message happens to be a particular string of ten single-digit numbers, then the alternatives to that number would be all the other possible ten-digit numbers. *Just-so-ness* stands out against a background of the alternative arrangements, the possible 'errors', that may have been constructed out of the sub-units of the message. Just-so-ness thus has a quantitative aspect; the more possible 'errors' there might be, the greater the just-so fidelity of the message.

The quantity of information in a message can also be related to the degree of surprise one experiences on hearing the message. News has to be unanticipated; old news is no news. Thus the more alternative messages there could be, the more surprising the just-so-ness of the actual message. Take, for example, the ringing of your telephone. To help illustrate Shannon's concept of information, we shall limit the message in our example to the identity of the caller, and not to the complexity of the ensuing conversation. This simplified message is transmitted when we pick up the receiver and the caller announces his or her identity by voice or word. Now imagine that your telephone is connected to only one other telephone, let's say to your mother's. In this situation, a ring can hold no surprises; it has to be your mother calling.

Imagine, in contrast, that your telephone is connectable, as it probably is, to all the other telephones in the world. A ring of the telephone now could herald anyone – your mother, the prime minister, a wrong number. In this case, you cannot know who's calling without picking up the receiver. The uncertainty of the caller is resolved only by answering the call. The revelation of the caller's identity is now true information. The caller is a *just-so* person, and no other.

Note that the *degree* of surprise varies with the caller. A close friend or relative may call regularly, and such a caller would cause you less surprise than would a call from the President of the United States of America (unless you happen to be the Secretary of Defense). Thus, different potential callers vary in the probability of their actually phoning you up. The world of alternative messages is a totality of different probabilities.

Shannon's insight was that any discrete 'message' (the particular caller in our example) bears an amount of information relative to the possible number of alternative arrangements (the numbers of potential callers in our example). Shannon formulated information as a *probability*. Given the universe of all possible callers, the content of information inherent in the identity of any one of them is the probability of that just-so arrangement. Information, therefore, is a *particular* arrangement relative to all other *possible* arrangements of the elements that make up the 'message'.

I put the word 'message' in quotes because Shannon's concept of information can be generalized to any collection of sub-units, irrespective of whether anyone or anything 'intended' the collection to be read as a message. One may compute the information present in a crystal of sugar molecules, a pattern of clouds, a protein or DNA sequence, a face, or a sonata, as well as in a written or spoken message. That is the power of Shannon's formulation, and that is why it has triggered a revolution both in concept and in technology far beyond the confines of the telephone call (§31, §32, §45, §58, §72).

Formally, Shannon proposed that a message's content of information was related to the logarithm (to the base 2) of the improbability (the *inverse* of the probability) of the message. The $\log_2$ formulates the information as a series of yes/no operations; indeed, this is the way information is encoded as *bits* in the standard digital computer program (0 or 1). This is also the strategy of common yes/no guessing games, like 'twenty questions'. The minimum number of yes/no questions theoretically required to arrive at the correct answer (a description of the message) is the amount of Shannon-type information in the message.

We can represent a message's content of information as the $\log_2$ of its intrinsic improbability:

$$\text{information} = -\log_2 (\text{probability})$$

In more formal terms, it looks like this:

$$H = -\sum_{i-1}^{n} P_i \log (P_i)$$

In a set of n possible symbols, the average information content per symbol (H) is related to the sum of the $\log_2$ of the improbabilities ($\frac{1}{P_i}$) of each possible symbol, each weighted by its probability (P_i). Information, in this formulation, can be seen to be the resolution of uncertainty. If, as in our first example, your telephone is connected only to your mother's telephone, then the probability that your mother is the caller is 1 (100%). There is no uncertainty to be resolved by picking up the receiver. According to Shannon's formula, the identity of such a caller bears no information: the log of 1 equals 0. (What the caller, your mother, has to say, of course, may be filled with information, and meaning, too. The information in a conversation can be computed by the probabilities of word usage, which is a very complex matter beyond our present concerns.) The information content inherent in a ring of a telephone potentially connectable to the world of telephones depends on the improbability of that person actually calling you.

The measure of uncertainty whose resolution generates information was called 'entropy' by Shannon. The word 'entropy', which is derived from Greek and means *transformation*, was borrowed by Shannon from the field of physics. Entropy in physics designates disorder and uncertainty in the physical world. Entropy is the *lack* of just-so-ness. Shannon adopted the term 'entropy' because a similar entropy equation can be used to describe a lack of a particular arrangement in messages, as well as in physical systems (§9).

At first glance, it might seem contradictory to use an entropy equation to quantify the just-so arrangement of a message. Did I not say that entropy is the *lack* of just-so-ness? Shannon's formulation becomes more reasonable, however, if we keep in mind that he defines a message as a vehicle for resolving uncertainty. The more uncertainty there is (the more potential callers there are), the greater is the information one obtains by resolving the uncertainty (picking up the ringing telephone). The just-so-ness in the message is equal in amount to the entropy, the uncertainty, waiting to be dispelled by the message. The numerical value of a just-so arrangement, a message, weighs exactly the same as the numerical disarray of the alternatives to the message, the entropy. Shannon merely uses the entropy, which is readily measurable, to replace its opposite, the information.

Do you think that Shannon has solved the problem of defining information, or has he only dodged the question? Does the internal arrangement of a message really suffice to define its content of information? Note that our ability to detect and quantify the information requires a background of knowledge that is extrinsic to the particular message before us. We need information beyond the internal arrangement of elements in the actual message itself. For example, to compute the entropy equation, we need to know the total set of symbols available in the alphabet used to deliver the message, and the probabilities expressed by each symbol. To know the probability of each of the symbols comprising the message, we need to have seen other messages in the past written in the same alphabet. A linguist, for example, could never reconstruct a dead language out of a single surviving fragment. He or she needs to see a large sample of messages to draw conclusions about the system of symbols and their rules of usage.

By equating a quantity of information with its improbability, Shannon solved the measurement problem for the information industry. Other problems, however, remain. Randomness itself is one of the problems. Any single sample in a random set is equally improbable, compared to all the other possibilities. The roulette wheel is a painful example. The number you choose, at random or by design, is improbable; someone else almost always wins. In fact anyone who makes a bet is unlikely to win. The winning number, although random, is very improbable. Is any random number the Shannon-equivalent of a real message? How can we distinguish a real just-so arrangement from a random roulette number, or random string of letters? The entropy equation cannot help here. Moreover, since Shannon-type information is not dependent on meaning, we cannot use meaning to distinguish a real message from a random sequence of symbols.

Nevertheless, experience can help us. The recurrence of an improbable just-so-ness argues against randomness. Repetition tells us something. If, for example, the same number were to come up consistently on the roulette wheel, or if the same person were to win time after time, we would surely suspect a fix. We might agree with Shannon that information is independent of semantics, that is, independent of meaning. Shannon is also correct when he equates an amount of information with its degree of just-so-ness. However, our ability to compute information depends on foreknowledge of the system of symbols used in the message. Moreover, our sense of information is dependent on our trust in the repeatability of the message. Indeed, the test of a scientific observation is its confirmation by others; an observation becomes a fact when the result of the experiment, however improbable, is publicly repeatable. A single result might be a random fluke. Real information is an arrangement reproduced.

§8 Information Impact

We have defined information; let us consider *meaning*, the child of information. The concepts of information and meaning are first-degree relations, but differ in generation. Information, according to Shannon, is an intrinsic order. But information that causes some effect is information that bears meaning. Meaning is the impact of information. Meaning, in contrast to information, is extrinsic. Meaning is what the information does. I have mentioned meaning here to help clarify a boundary of information, to tell us what information is not. Below we shall discuss meaning in greater detail (§66).

§9 Energy and Order

Let us return to causality. What do we mean when we say that causality may be ascribed to energy transfers (§6)? What is energy and how may it be transferred?

The word 'energy', like the word 'information', is widely used; and its meaning, too, is considered to be obvious. Most physics texts define energy simply by using the term in equations describing the basic laws of physics. If physicists bother to define energy using words, they may say that energy is the capacity to do work. Indeed, the word 'energy' comes from a Greek word meaning work. Work implies constructive action; but energy, as everybody knows, can also be explosively destructive. Energy can assume different forms manifest in heat, chemical bonds, atomic nuclei, momentum, electromagnetic radiation and other moving waves. Energy may or may not be felt within us when we get up in the morning, and move. Common sense associates energy with motion; energy is expressed by matter (or a wave form) that moves, or that could move, potentially.

The transformations of energy in its various forms are described in a discipline termed 'thermodynamics'. Historically, thermodynamics arose from a study of the flow of heat (thermal) energy, but its principles are applicable to all transformations of energy.

Technology can fuel science. Theories of information emerged from the communications revolution of the twentieth century, and thermodynamics emerged from the industrial revolution of the nineteenth century. The exploitation of energy on an industrial scale inspired people to study steam engines with scientific fervor. The 'laws' of thermodynamics were discovered in the process. I wrote 'laws' in quotation marks, because thermodynamics describes *principles* of nature that apply in a certain condition. By contrast, a classical law of nature, such as the law of gravity, applies unconditionally. The distinction, however, should not mislead us; we are ruled entirely by the principles of thermodynamics, and so we shall refer to them as laws, without quotation marks.

What is that special condition that admits the laws of thermodynamics? That condition is deceptively simple; the laws of thermodynamics relate to *closed* systems. We shall discuss more about systems below (§45), but here we can define a closed system by analogy to an ideal closed room: a closed system is a physical entity sealed off from all outside influences. No energy or matter can enter a closed system, our closed room, from the outside. Anything that happens, happens by the power of what's already there.

The first law of thermodynamics says that energy is conserved. Energy may be transformed from one form into another, and it can flow from one particle of matter to another; but energy cannot be destroyed. Even energy that has left the system, is lost but not destroyed. (Note, for energy to be lost, the system has to be *open*.) Since energy cannot be destroyed, energy has consequences. The nail must respond to the blow of the hammer. Energy thus can act causally.

The second law of thermodynamics tells us two things essential for causality: first, that energy can flow spontaneously; and second, that it flows spontaneously in a particular direction. It is the spontaneous flow of energy that makes life (and our world) possible. If energy were not to flow, there would be no chemistry, and hence no life. Everything would be frozen in place. Nothing would move, unless some outside source of energy were to enter our closed system and move things. The spontaneous flow of energy is the internal force that fashions our world.

The second law of thermodynamics also tells us that in every transformation of energy, some of the energy flows spontaneously in one direction only, into disorder. Disorder, obviously, is a lack of order. But what is order? Order is akin to information, as we defined it above (§7). Order is a particular arrangement of a situation distinguished from all other possible arrangements. The flow of heat provides a familiar example of the flow of energy into disorder. Consider a hot bowl of soup in a cool room (remember, the room is closed). The heat energy occupying the room shows a very particular arrangement: it is concentrated in the soup. The arrangement of the heat is ordered.

The soup is a point of hot energy relative to the uniform distribution of heat in the rest of the room.

Common experience tells you what will happen in time to the hot soup; the soup cools, allowing you to eat it. The inequality in the concentration of heat energy, an ordered arrangement, will disappear, *spontaneously*. There is no need for any intervention. The heat flows out of the bowl, of itself. The flow continues until the soup reaches 'room temperature', a temperature equal to that of the rest of the room. Actually, the soup, as it cools, heats the room. The heat gained by the large room from the small bowl may be imperceptible, but it is still there. Remember the first law: energy cannot be lost from the room. The heat has only scattered from its concentrated arrangement in the bowl of soup to a homogeneous dispersion throughout the room.

Of course, the flow of heat energy stops once the temperature of the soup and the room are homogeneous; there is no longer a cold 'sink' into which the energy of the soup might flow. The flow of heat energy stops at *equilibrium*. Individual molecules in the soup or air may move with more or less than average speed (more or less energy), but once equilibrium is attained, the system maintains its average homogeneity. Because the initially ordered concentration of heat in the room has disappeared, the spontaneous dissipation of the heat from the soup can be viewed as a flow towards equilibrium which, from this point of view, is disorder.

Note that the spontaneous flow of energy is one way only. Cold soup has never been seen spontaneously to extract heat from cold surroundings and warm up. Heat always flows from the warmer to the cooler, never from the cooler to the warmer. Thus, energy flows by its very nature, and it flows in the direction of greatest probability, that is from more order (arrangement) to less order (homogeneity). Indeed, if you want your soup to stay warm, you will have to keep the fire going by supplying more energy. Energy spontaneously dissipates. That, in a nutshell, is one of the expressions of the second law of thermodynamics. Amazingly, as we shall soon see, the natural dissipation of energy fosters causality.

Why, indeed, does energy move into disorder? Logically, it is obvious that situations of greater probability are more usual than are situations of lesser probability. That, in fact, is what probability means. In any given collection of items, there are fewer ways to create a particular arrangement (order) than there are ways to annul that arrangement (homogeneity, disorder); recall our discussion of information (§7). Thus, disorder is far more probable than is order. Hence, disorder will tend to increase spontaneously in a closed system. Entropy is another name for probable disorder, and entropy is defined by an equation very similar to the one used by Shannon as a measure of information (§7). Shannon, in fact, borrowed entropy from thermodynamics to describe the improbable arrangement of information.

Physically, we can view the dissipation of energy as an intrinsic property of some forms of matter. Energy, we said above, implies motion. Molecules bounce into each other and atoms vibrate, unceasingly. Indeed, the third law of thermodynamics says that it is practically impossible to freeze molecules into perfect stillness (the temperature of absolute zero). Since momentum is conserved (the first law), a fast-moving molecule that bumps into a slow-moving molecule will transfer some of its energy. Hence, the fast molecule will slow down and the slow molecule will speed up. As a result of their incessant bouncing about, collections of molecules in contact will end up moving with a constant average speed. The system ends up in equilibrium: that's why the soup cools while raising the temperature of the surrounding (closed) room. Energy averages out. Indeed, the more energy there is in a system, the more unstable bouncing there is, and the faster the system will tend toward disorder. Chemical reactions, therefore, tend to progress to the states of lowest potential energy possible *under the circumstances*. Different circumstances may dictate different *states of lowest potential energy*. Nevertheless, as we shall see, the relative stability of systems at low states of potential energy is critical to the chemistry of life (§101).

The second law of thermodynamics, in summary, says that energy spontaneously flows from coherence, an improbable orderliness, to incoherence, a random probability. Entropy, disorder, constantly increases.

If this is so, how is it that we see so much order in the world? How can a just-so structure arise from the dissipation of energy? Work requires order, coherent motion; how can coherent work emerge from incoherent energy? The answer is that order can emerge by the conjunction of two factors: by opening the closed system and supplying more energy from the outside, and by contriving to trap the incoming energy. Common experience teaches that order requires investment. Spontaneously, desks can only become messy, never orderly; to put a desk back in order demands hard work, the addition of more energy. Warming the soup, as we said, requires operating the burner. Even though a significant part of the invested energy also tends to dissipate into entropy, the entropy, paradoxically, can be harnessed. The spontaneous flow of energy into disorder supplies a motive force that may be exploited by some contrivance.

What kind of a contrivance can harvest energy? Consider a sailboat in the wind. The wind is the epitome of disorder. The explosive self-destruction of the sun hurls energy into space, some of which chances to hit the earth. The different reflectivity and absorbency of the land and water scattered over the earth's surface produce local differences in temperature that heat or cool the overlying air. Differences in heat lead to differences in atmospheric pressure, and the spontaneous flow of heat and pressure towards homogeneity, described by the second law of thermodynamics, is felt as the ever-changing weather and the force of the blowing wind. What is more chaotic than the weather, or more unpredictable than the wind (§25)? Yet, out of such fundamental disorder the experienced sailor can navigate his boat and its cargo to port. The haphazard wind is put to work by the sail, the mast, the boom, the rope, the keel and the rudder, all in combination with the will and skill of the sailor. The sailor can contrive to use the wind's energy to move his boat in the direction of his choice, and, by tacking, even against the direction of the wind. He need only adjust the sails and rudder accordingly. The only condition for working the boat is that the wind blow. No wind, no sailing. In other words, the chaotic dissipation of energy by the wind is required for the orderly sailing of the boat. Thus, the very flow of atmospheric energy into disorder provides the sailor with the causal force he needs to sail his boat in an orderly fashion. Without the disorder, which is the wind, there can be no order, which is the boat's course. The second law of thermodynamics describes the irrevocable progression of nature into disorder, the inevitable increase in entropy. Yet, contrivances such as sailors and sails can harness the disorderly flow to create order. Ship-builders, sail-makers and sailors just have to expend energy and work at it. The sailor and the boat are a system open to influxes of energy, some of which is the wind.

Note that even while the sailor, the sails and the boat harness the wind, all three are themselves progressing into disorder. The sailor tires and ages, to be replaced one day by his son, while the boat rots beneath him and the sail frays above him. The sailor eats and metabolizes food to extract the energy he needs to carry on. The life and growth of plants and animals is cut short and destroyed to afford the sailor his hour at sea. The sailor is an open system, who lives by extracting energy from the world. Disordered energy thus can be transformed, at least in part, into order, provided that the system can exploit a constant flow of new energy. Remember, no wind, no sailing.

Above, I said that the concept of a closed system was 'deceptively simple'. In reality, no system is closed absolutely. Heat flows, however slowly, through the best insulation. Your cellular telephone operates in the deepest basement. Cosmic rays bombard the earth from the farthest reaches of space. Life, which can exist only in open systems, exists the world over. Perhaps only the Universe itself is a closed system; nothing material exists beyond its 'borders'. (But, we are told, the Universe is continually expanding; perhaps even it is not closed.)

Now if the Universe is a closed system, then thermodynamics teaches us a very important fact about the Universe. The second law says that entropy must incessantly increase in the Universe as a whole, despite any local or temporary emergence of order. Let us return to the wind and the sea; the destruction of foodstuffs and raw materials needed to build and maintain the sailor and the boat is costly. Thus, the increase in the sum total of disorder added to the Universe more than offsets the local creation of order that is the sailor, the boat and the course they take.

The second law also tells us that no contrivance, however efficient, can convert energy entirely into coherent work. There must always be waste, residual disorder, lost heat or friction. Only a fraction of any energy is 'free energy' that can be harnessed for work. In effect, perpetual motion machines are impossible, or, if you prefer the vernacular: there ain't no free lunch. Evolution, as we shall discuss (§31), is a contrivance for harnessing the energy of the dying sun to create, among other things, sailors. Human culture (§39, §79) is a contrivance for harnessing dissipating energy to fashion, among other things, boats. Contrivances that harness entropy, like the sailor and the ship, ultimately rot into death and disorder. Fortunately, the sailor dissipates at a slower timescale than does the wind.

§10 Causal Information

To return to causality, we can classify constructive and destructive energy transfers according to whether they preserve or increase information (maintaining or building order) as opposed to their wasting information (dismantling order into randomness). Molecular signaling is a transfer of information, as is visual and auditory communication. Perhaps the most telling example of information transfer in biology is genetics: the transfer of information from the nucleic acid sequence of DNA into the amino acid sequence of proteins. The amino acid sequence, in turn, causes proteins to fold into specific shapes (structures). Protein structure then gives rise to the specific functions of the protein (§101). Thus, biologists see a 'chain of causality' progressing upwards, as it were, from information (DNA) to structure (protein) to function (enzymatic catalysis, antibody recognition). Scientific reduction, in a sense, is like going down the chain of causality; biologists strive to reduce the complex functions of living systems to their genes, a reduction that takes evolution into account (§28, §29). Below, we shall see that such reduction is not so simple (§11).

Note that the chain of information transfer in living systems is driven by a chain of energy transfer, metabolism (§47). It takes hard work to maintain information, just-so-ness, in the face of spontaneously increasing disorganization, entropy. Information and constructive energy have to proceed hand-in-hand. If, as I said above (§9), information is a *reproducible* arrangement, then the reader might consider adding to Shannon's entropy equation a measure of the amount of *effort*, additional information, needed to reproduce the message.

§11 Limits of reduction

The reduction of complex phenomena to underlying causes (laws of nature, agents, internal structure, energy, information), is the essence of science. Reductionism is authoritative, but it is not a panacea. Reduction is suitable when the phenomenon in question can be boiled down to the properties of an underlying physical unit, a material substance such as a particle of matter, or a biological entity such as a bacterium, a protein molecule, or a gene sequence. Infections, for example, are adequately reducible to the infectious agent, catalysis to a protein, an inherited disease to a faulty gene. Reduction is the replacement of a superficial appearance by a deep causal reality; the phenomenon of interest is demonstrated, with the aid of experimentation, to be inherent in the character of the sub-units that constitute the subject of study. Successful reduction, in essence, crosses scales; it discovers microscopic entities or processes whose very existence *necessitates* the macroscopic phenomenon of interest.

The process of reduction does not complete the scientific enterprise, however, when the subject in question is not a discrete entity but rather a relationship or an interaction. Here, the subject is an organization. The individual characters of the system's sub-units may not be able to account for the behavior of the system when the system is complex. Immunologists have been shocked, as we shall see, to discover that the immune system of a mouse may go about its business as usual, despite the removal of key immune genes from the DNA of the mouse (§129). The immune system can apparently learn to organize its behavior using different sets of sub-unit molecules. A particular gene may be essential, but only after the system has organized itself around that gene, has come to depend on it. Conceivably, the system could have organized itself around a different gene, that alternative gene now becoming essential. The learning process characterizes self-organizing systems at various scales (§71). Is a happy life contingent on connecting with the one *essentially right* mate, or could the person have organized an equally good life with a different *essentially right* mate? Literature continues to explore the issue of *essential* human relationships. But *essential* molecular relationships too may show functional plasticity. The point is that organizational entities depend on the interactions of their sub-units, and not only on the distinct characters of particular sub-units. The individual sub-units by themselves may not contain the essence of the organizational entity. Delving beneath appearances into the microscopic sub-unit scale, of itself, cannot provide complete understanding. The essence of organizational entities emerges from the interactions of the sub-units. Some scientists may feel that the study of emergence is not real science, that the soul of science is reduction alone; others may feel that the study of emergence is the natural extension of reduction. Read on and form your own opinion.

EMERGENCE

Reduction, however useful to the pursuit of science, does not always suffice to explain important properties of many complex systems, particularly life. The causal concepts of emergence and of attractors are necessary to complement reduction in a scientific account of our world.

§12 Emergence

Emergent properties can be defined differently according to different interests. Physicists might look at the solid, fluid and gaseous phases of water as emergent properties of aggregates of water molecules in different energy states. Here I want to focus on complex entities that emerge from the interactions of heterogeneous elements. The emergent properties are not inherent in any of the elements of the system examined in isolation. The interactions themselves generate properties that characterize the system as a whole.

Subjects defined by emergent properties include the most interesting questions of biology. Consciousness, speech and emotion are not housed in the brain as discrete physical entities; these aspects of mind emerge from the way neurons are functionally organized in the brain. You cannot productively study speech by studying isolated neurons. Isolating the neurons, in fact, destroys the phenomenon of speech. Even more mundane, manifestly physical performances of the brain are emergent entities, including walking or shaking hands.

Ecological niches, the accommodations of living creatures to their special environments, exist as organized interactions. Reducing the niche to either the creature or the environment in isolation destroys the essence of the niche. Evolution itself, the motive force of life, is an emergent phenomenon.

Economies, entities which are objectively quantitative, emerge from the interactions of people, materials and media of exchange. Economies are not reducible to their components; they cannot be explained by analyzing collectives of people, materials or coins in isolation. Economies exist as interactions and are susceptible to study as interactions.

Entities in the material world too emerge as interactions. A river, for example, cannot be understood by reducing it to molecules of water. The river exists as an interaction between a flow of water, the materials making up the river bank and channel, the elevation and terrain of the land, the climate, temperature and rainfall, the law of gravity, the forces of turbulence, and more. Component parts of the river are reducible, one by one. The river as a totality, however, is not reducible to any list of its microscopic sub-units. The river is not inherent in any of its constituents; the river is better understood as a dynamic relationship between its sub-units. The river, nevertheless, is real. We can dam it, fish it, make a living from it and, if we don't know how to swim, drown in it.

Heraclitus, an early Greek philosopher, was reported to have said, 'You never step into the same river twice.' This statement is the gist of the argument. The river reduced to a flowing collection of water molecules never, in truth, contains the same collection. Your underlying molecular constitution is no less in constant flux. Nevertheless, defined as emergent systems, both you and the river remain the 'same', the flux of molecules notwithstanding.

§13 Emergent Life

The clearest example of an emergent property is life itself. Life is not inherent in any single element constituting the living cell. DNA is not alive, neither are proteins, carbohydrates or lipids. Indeed, for a single short moment, a living cell and a dead cell may, upon analysis, be found to contain precisely the same catalogue of 'dead' chemicals in identical concentrations. Bacteria have been resurrected after 35 million years of suspended life in the guts of ancient bees entrapped in amber. While not quite dinosaurs, 35 million-year-old bacteria are still a marvel. Today they surely live; what was their state for 35 million years? What distinguishes the living from the dead? Nothing more than actions and interactions. Life emerges from inert matter as a consequence of metabolism, the continuous transfer of energy and information systematically packaged in cells in a way that leads to self-perpetuation (§47). The complexity of dynamic behavior that generates metabolism, growth and genetic inheritance is what we call life. In that piece of amber there existed, lifelessly for 35 million years, all the components needed for life, but there was no life until the machinery actually began to interact. Dead molecules metabolized the bacteria into life.

The emergent property of life is so far removed, so foreign to the material sub-units comprising life, that life seemed to demand the addition to matter of a special 'vital ingredient'. It was inconceivable to our forefathers that the quantitative connections of matter could ever cross the qualitative chasm that separates the animate from the inanimate. Living bodies had to be animated by a 'breath of life'. Today, Western science can explain the emergence of life from chemical principles without recourse to mystical vital powers. But that's the wonder of emergence, the generation of wonderfully new qualities from the behaviors of material systems.

§14 Understanding Emergence

The study of emergence and emergent systems cannot be done adequately by reduction alone. It is equally true, however, that emergence cannot be studied without reduction. To be able to understand and manipulate an emergent system, we have to analyze it; we have first to break it down into each of its constituent parts to see what it's made of, to see what the system emerges *from*. But we can't stop at reductive analysis: we have to continue and put the system back together again. Conceptually, or better by quantitative modeling, we reconnect the interactions between the constituent sub-units. Then we can begin to see how the properties of interest emerge from a dynamic organization. Only by combining analysis with synthesis, do we achieve useful knowledge. If, for example, we want to control a wild river, we first must analyze the flow, the turbulence, the characteristics of the river bank and the river channel, the terrain, the rainfall, and so forth. After we put it all back together and construct a dynamic model of the river, we can know how and where to set up the system of dams to do the most good at the least cost.

Likewise with emergent biological systems (like the immune system). First we have to analyze them, causally reduce them to their material sub-units (tissues, cells, molecules, atoms, interactions). Then we can connect the sub-units to synthesize a dynamic model. A composite understanding of the organization allows us to plan the safest and most effective way to intervene to obtain the desired result (the cure of an autoimmune disease, or the development of an effective vaccine, for example). For the past century, immunology has, with great success, been occupied in analyzing the immune system into its molecular building-blocks. The field is now ripe for synthesis. This book is one introduction.

§15 Multiscale Emergence

Note that complex emergent systems may contain independently emergent sub-units. An area of turbulence, for example, is a macroscopic sub-system of the river, which can be analyzed meaningfully in terms of the microscopic behaviors of collections of water molecules. Similarly, a living organism can be decomposed into separate organs, such as the brain or the immune system, which themselves are emergent sub-systems. Thus, emergence can be multiscale: small emergent systems organizing themselves together to form grand emergent systems. It is possible to break down grand emergent systems without destroying constituent emergent sub-systems; cells can be removed from the body and grown in test tubes. The human mind likes to analyze, break down complexity. Nature likes to go the other way; simple emergent systems synthesize grand emergent systems, cells to organisms to societies. That building process is the subject of evolution, to be discussed below (§28–§36).

§16 Terms of Emergence

In the following paragraphs, we shall enlarge our discussion of interactive systems using three terms to describe features important for emergence: state, attractor, and basin of attraction.

1. The *state* of a system refers to its activity.
2. The *attractor* of a system refers to its stability.
3. The *basin of attraction* refers to the way the system reacts to perturbations and disturbances.

In addition to these terms, two other properties of interactive systems are worth noting: time and control. Time is important at several scales. Biological systems, similar to other complex systems, are influenced by their past histories; they evolve. The importance of control is self-evident. These five elements form the thread of our discussion.

§17 System States

Emergent systems are dynamic; their constituent components and the connections between them change with time. Minds experience and learn, bodies grow and heal, immunity overcomes infections, rivers collect rains, economies survive wars. Changes in systems over time have been represented and studied mathematically in various ways: differential equations, geometrical or topological constructions, network configurations (neural nets, Boolean automata). These and other ways of studying system dynamics define or characterize the overall configurations of the system that can be called the *states* of the system. Minds can be in states of tranquillity or agitation, bodies in health or disease, nations at peace or war. The global state of the system emerges from the states of the individual components and how the components affect one another through their mutual connections. Nerve cells fire signals or don't fire signals, lymphocytes make antibodies or not, people have jobs or sit at home.

§18 Emergent Experiments

Keeping track of complex transitions over time and predicting the global states that will finally emerge is too complicated for most human brains. The behavior of dynamic emergent systems is often studied with the aid of computer simulations. Discrete 'cells' representing the sub-units of the system of interest are connected in various ways and made to interact according to rules programmed into the simulation, stimulating, inhibiting or modifying one another. The parameters and variables can be chosen to reflect or mimic aspects of the natural system of interest. The program is then run on the computer. By adjusting the interactions, the experimenter can obtain behaviors that faithfully replicate real features of the natural system of interest, be it a collection of heated molecules, a network of nerve cells, a colony of ants, an ecosystem, or an economy. The simulator can then perform experiments by perturbing the system, adding or changing components or rules as he or she wishes. Once we have set up a model of a dynamic, interactive system, we can let the computer run through the states of the system generated by the activities and connections of the components. In this way we can see what the system produces, what emerges.

The observed effect on the patterns of behavior of the system is an experimental result that can be tested in the real world, if that is desirable or feasible. Indeed, the very transformation of a real river into a useful dynamic model of a river, one may argue, proves that interactive, emergent systems may have an objective existence independent of the physical materials that constitute them. In other words, the phenomenon of emergence is a property of interactions as such, not a property of particular substances. Although the subject of immunology is biological, we can illustrate emergence using whatever analogies seem convenient. Emergence is immaterial. Of course, the concrete manifestations of the emerging interaction are dependent on the particular substances that interact. Because fluidity is a property of water, what flows in the dynamic system we call a river is the water. What lives are cells.

§19 Attractors

A striking property of some systems is that, over time, they can settle into a stable existence, a stable state. The emergence of stability is demonstrable logically in mathematical formulations, but long-term stability is also observable in natural systems. A river, for example, can remain the same river despite Heraclitus (§12) and the incessant flux of water molecules, the intermittency of the rain, and the cycle of the seasons.

Sometimes, the long-term outcome of a system can appear as an oscillation between states (day and night, for example) or as a cycle of states (spring, summer, fall, winter, for example). Some systems never settle down to any stable behavior, but wander for ever into unpredictable configurations. A comet that has not yet been entrapped into a fixed orbit by our sun or by any other star may, for example, wander for ever in the infinity of space in a haphazard journey buffeted by gravitational pulls and kicks. At a much smaller scale, we may imagine the unpredictable motions of single molecules wandering about in a gas unconfined.

The long-term stable state of a system is called an *attractor*. Just as a magnet will pull a random collection of powdered iron into a stable configuration of the iron around itself, an attractor will 'pull' the interactions of the dynamic system into a certain stable configuration of its elements. The river is the stable configuration, the attractor, generated by the interactions of the rushing water, the channel, the law of gravity and all the rest. The living cell is the attractor of all the metabolic and genetic interactions that form it. The wandering comet and the wandering gas molecules, uncaptured by stable interactions, are without attractors. Attractors are islands of coherent order in a sea of expanding incoherence, entropy (§9).

§20 Natural Attraction

The word 'attractor' is odd; it has a sense of final Aristotle-like causality. The interaction is attracted, pulled as it were, to its appointed state. However, the term 'attractor' was introduced by mathematicians, who may be forgiven. After all, the solution to an equation is inherent in the very formulation of the equation, even before the equation is solved. Equations are equivalencies, tautologies. The attractor, which is a solution to the dynamics of the system, exists *a priori*, even if the emergent attractor surprises the modeler who formulated the model. The attractor is inherent in the formulation; it need not be inherent in the formulator. Mathematically, attractors do attract.

Nature, however, is not mathematics. Natural systems are not equations devised *a priori*. Natural systems may or may not settle into an attractor, without anyone having designed them to do so. A natural system, we believe, is blind to its final state. The behavior of the system, rather than being attracted to stability, creates its own stability. Thus, one may argue that the word 'attractor' may be a misnomer for natural systems. Nevertheless, natural systems do attain long-term stability. Hence natural attractors are demonstrable, at least in a metaphorical sense. I shall apply the term 'attractor', therefore, to describe any relatively stable state of a natural system that manifests interesting properties. From whence comes such stability?

§21 Stability

An attractor is a global state that manifests long-term stability. Stability is an operational property; attractors are states of the system resistant to perturbations. The river, for example, maintains its customary configuration despite fluctuating rainfall, floods or droughts. Nations survive wars, bodies recover from illness, agitated minds find peace. Stability reigns, but only as long as the perturbations are not too drastic. A system can indeed be knocked out of its attractor, irreversibly. The annual flooding of the Nile, which over millennia molded a civilization, was stopped by the Aswan High Dam. An asteroid striking the earth triggered the mass extinction of the dinosaurs 60 million years ago, and made space for the mammals to take over (§31). An HIV infection can terminate an immune system and a life.

Being evicted from one attractor, however, does not necessarily lead to the disorderly end of a system. A system may settle into another attractor (§30), another form of order (§9). A river can change its channel, and even its direction of flow in the wake of a shift in the terrain after a volcanic eruption or an earthquake (§28). Leaving the farm, a farmer may become an industrialist. An ageing athlete may settle down to being a sportscaster. Birds may evolve to lose the power of flight and take to the sea (penguins) or become runners (ostriches).

Not all perturbations have to be as energetically extreme as are stellar impacts, high dams, social upheavals, epidemics or speciation to influence the long-term state of a system. A teaspoon of penicillin can eradicate an infection that would otherwise lead the patient to a death attractor. A single word, a glance, can change the course of a lifetime. An idea can change the course of the world. Thus, the power of a perturbation is not proportionate to the magnitude of its energy. Rather, the impact results from the sensitivity of the system to the discrete information or energy that perturbs. In other words, systems can be intrinsically susceptible to lasting modifications by selected inputs at critical states or moments. Our knowledge of the nature and timing of critical inputs, as we shall see, provides us with control, the power to make changes. The sensitivity of a system to control is related to the third attribute of an interactive system, the system's basin of attraction.

§22 Basins of Attraction

Using a geographical or topological metaphor, attractors, like rivers, can be thought of as fed by basins of attraction. A system's basin of attraction is the set of perturbations whose impacts will still let the system return to its normal attractor, its historical channel. A system's basin of attraction is like the landscape that supports a river. All the rain that falls on the river side of the hills, the watershed, flows into the river.

Beyond the historical basin of attraction, beyond the hills bounding the river valley, lie other basins that can attract the system into new valleys. The rain that falls to the west of Jerusalem flows to life in the Mediterranean Sea; the rain that falls to the east of Jerusalem enters a basin of flow whose attractor is the bitter Dead Sea. The farmer who moves to the city enters a new economic basin of attraction. The green wave of traffic lights is a basin of attraction conducive to flowing traffic; grid-lock is the attractor when the traffic lights go out. An attractor of the healthy heart is its normal pulsation; a coronary occlusion can push the heart muscle into a basin of attraction leading to arrhythmia or standstill. Luckily, return is possible. The resuscitation team may appear just in time, as we have learned from TV, to administer the right electrical stimulation needed to knock the heart back into its healthy rhythm.

§23 Multiscale Attraction

Just as grand emergent systems (like a living organism) may be composed of emergent sub-systems (like cells and organs), so may more complex attractors harbor sub-systems, each sub-system participating in its own attractor. An economy, for example, may operate as a globally stable state (attractor 1); the people who participate in that economy exist each in his or her own state of stable activity (attractor 2); the organs (attractor 3) and cells (attractor 4) forming each person can be seen to operate in their own dynamically stable states. In other words, the world, like a Russian doll, is filled with attractors embedded within other attractors occupying diverse scales of space and time.

Note that nested natural systems tend to be mutually connected such that their states of stability, their attractors, are functionally dependent. The collapse of an economy may lead to a person's loss of gainful occupation and, ultimately, to the person's death. Dependence between scales of organization can also go from the smaller to the greater. A virus epidemic, for example, may kill enough cells, organs and then people to undermine an entire economy; measles and smallpox, and not only the immigration of Europeans, wiped out whole cultures of Amerindians. Unlike the defined attractors of mathematical equations, the attractors of natural systems are open, compounded of contingent sub-systems and, therefore, can be creatively complex.

The interdependency of natural systems and sub-systems can generate conceptual paradoxes. Our farmer who moves to the city and becomes a shopkeeper creates a definitional dilemma. Viewed as a person, Mr Jones is the same Mr Jones, the same attractor, irrespective of whether he feeds chickens with corn on the farm or feeds people with chickens in the city; just ask Mrs Jones. However, viewed as a discreet element in a larger economic system, farmer Jones is not the same sub-system as is shopkeeper Jones. In his person, Jones is one stable attractor; in his changed economic role, City Jones participates in a different attractor than did Country Jones. Since attractors designate properties of whole systems, we should define the borders of the particular system when discussing attractors. But a problem with natural systems is their fuzzy borders. All are interconnected.

§24 Control

A control element can be viewed as the discrete information or energy needed to maintain a particular set of interactions, a basin of attraction. A change in a control element can push a system into new attractors, new states of long-term stability. Metaphorically, we might say that a control element is like a dip in the landscape of the hills, the watershed that separates two valleys. A control element can act like a tunnel or siphon to connect the river valleys beneath or over the hills. It is relatively easy, it requires less energy at such critical connection points to move the river water from one valley to the next. The basin of attraction is thus a topological metaphor for the conditions that delineate the stability of a system's behavior.

Each system has its characteristic topology of sensitivities. An attitude may be changed by some words but not by others; a bacterium may be destroyed by one antibiotic but not by another; an autoimmune disease may be triggered by administering a certain drug to one person, but not to another. A system's natural control point is a point in the basin of attraction at which the system is sensitive to perturbations that can change the system's attractor. If we want the system to persist in its behavior, then we should avoid the critical dips, the sensitive points, in the basin of attraction. If, on the contrary, we want to cause a change in the system, then we should direct our efforts to its sensitive spots. Each system has its own setup, its own strengths and weaknesses, its own topology of attraction, its own borders of order and disorder. Knowledge of the landscape can provide control. We'll talk more about control below (§26), and when we discuss health and disease (§166).

§25 Strange Attraction

We have discussed three kinds of outcomes that characterize the behaviors of interactive systems: stable attractors, attractors that feature oscillations (like day and night), and the lack of any attractor (like the unpredictable path of the wandering comet lost in space). To be complete, we should also mention a fourth behavior: the strange attractors of chaotic systems. Strange attractors were first born as creatures of certain mathematical formulations. Strange attractors are computational outcomes which show a very high sensitivity to initial conditions. In particular mathematical iterations, for example, infinitesimally small differences between starting values can lead to large differences in outcome that only enlarge with time. Chaotic systems, as it were, move in bizarre basins of attraction; even the smallest perturbations count. Another distinctive feature of chaotic systems, in fact the reason we call them chaotic, is that changes in the state of such systems never precisely repeat themselves.

An everyday example of a natural chaotic system is the weather. Although the weather people have detailed knowledge of the global influx and outflux of the energy that drives the weather, can consult satellite pictures of the world's weather at any given moment, have in hand maps of temperatures, pressures and flows, and own super computers to put it all together, these professionals are unable to forecast the weather with accuracy more than four, perhaps five days in advance. Their honor, nevertheless, remains intact; the weather system is officially chaotic. Fortunately for them, even Athena herself, the all-knowing goddess of science, could do no better. Fortunately for the world, chaotic systems are not totally random; they can manifest regular patterns on a coarse scale. Beyond the equator, we do have seasons, and winters are indeed colder than are summers. Unfortunately, the term 'chaotic', at least to the ears of the uninitiated, has the ring of the random. The lay public seems to have been charmed by the idea that chaos may now be under control, but mathematically, chaos and randomness are distinct from one another.

Are there other natural systems, in addition to the weather, that feature strange attractors? It is difficult to judge. A system that produces non-periodic oscillations could be a system truly caught in a strange attractor; it could also be a truly random system. Moreover, non-periodic behavior could result from the buffeting of an otherwise stable system by 'noise'. Indeed, mathematical modeling may not always resolve the attractor classification of natural systems. Therefore, the distinction between a 'regular' attractor modified by 'noise' and a truly strange attractor with coarse stability, although fascinating in concept, is not practically important in our present discussion and I mention it only in passing. The important thing is that natural systems can manifest long-term stability at certain scales. Such a property, by convention, is called an attractor.

§26 Attraction Control

The tuning of attractors can be a helpful way to think about control. Mathematically, a given set of variables can generate a standard stable attractor, oscillations, a strange attractor, or no attractor at all, depending on the parameters associated with the variables. (The *variable* in a system represents the 'unknown' whose character is influenced by associated factors called *parameters*, which we may adjust.) The consequences of playing with parameters may be extreme in natural systems. Consider a set of interacting people, a family of parents and children or a triangle of lovers, for example. Each person can be considered as a variable capable of different states of mind and behavior. The connections between the people, their relationships, are akin to adjustable parameters. Each of us can attest to the violence or harmony that may be visited upon the family or the lovers following what appear to the outsider to be minor changes in some relationship connections. Harmonious control or disruptive loss of control can result from minimal adjustments in the parameters of interaction between the people. The variables we choose for the immune system, of course, will be cells, not people, and the parameters that define their interactions will be molecules, not feelings. Yet we shall see how the same family of cells and molecules may produce either health or disease depending on the nature of their signal connection (§160). We sometimes can get desirable results by adjusting the connections (family therapy), without necessarily having to dismantle the system (divorce). Such are the wonders of emergence.

§27 Mathematical Law

It is a wonder that nature, in some of her guises, actually seems to behave according to mathematical law. The second law of thermodynamics is more a statistical description than it is a physical law (§9). The concordance between mathematics, an invention of the human mind, and fundamental laws of nature such as gravity amazed Einstein. Perhaps more amazing is the existence in nature of attractors: stabilities of awesome complexity that yet fit a mathematical formalism. The mathematical rigor of nature, however, did not amaze the Greeks; mathematics is included in Aristotle's description of cause, the formal cause (§5, §77). To Aristotle, form, including mathematics and logic, is given by nature to us, not imposed by us on her. Some biologists, immersed in their reductions, may wish to ignore emergence. Nevertheless, biology is a science that aims to understand and control emergent phenomena. Biology, the science of life, rides on ultimate emergence. Mathematicians and physicists who study complex systems are beginning to work with biologists to discover the fundamental laws that rule complexity.

EVOLUTION

Evolution is the formative principle of life in its varied manifestations, including immunity. Exploring the seminal concepts of adaptation and of fitness, we shall see that we might more accurately account for the process of evolution using the concepts of attractors and of fittedness.

§28 Time and the River

The word 'evolution' comes from the Latin *evolvare* (*volvo*, like the name of the car, means 'I roll'), and evolution means to roll forth, to unfold. Evolution refers to the emergence of change over time. Evolutionary change embraces two, seemingly opposite, concepts: progress and stability. Evolutionary change is stable because it perpetuates itself, and yet it progresses because it develops adaptively. (Self-perpetuation and adaptation will be explored in more detail as we proceed; §30.) The idea of evolution has driven much of biological thinking ever since Darwin proposed natural selection as the mechanism responsible for the development of living species. Darwin explained that species, breeding populations, change over time because some variant individuals transmit their characteristics more effectively than do other individuals to future populations. The dominant variants are those individuals who have survived the trials of life better than their standard fellows (have undergone *selection*), and who have succeeded in passing on their successful characteristics to their likewise successful offspring. The once variant, over time, becomes the standard; the population has been transformed by selection, has evolved.

Darwin's associates, T. H. Huxley and others, were quick to see that processes of natural selection might explain, not merely the emergence of living species, but the transformations of natural phenomena *universally*. *Variation*, *selection* and *perpetuation*, the triad of Darwinian evolution, can be seen in many natural complex systems. (For a readable exploration of the idea of universal Darwinism, the reader is directed to *Darwin Machines and the Nature of Knowledge* by Henry Plotkin [1995].)

The process of evolution, an emergent property of life, is critical to immunity on several scales. Some components of the immune system have evolved over millions of years, while other components evolve over the decades of one's lifetime, and yet others evolve during the hours and days of individual immune responses. Thus immune evolution occupies widely different timescales. The details will be discussed below, but there are some general points about evolution that relate to causality. One point is that biologic evolution, like other emergent systems, records a history. What has occurred in the past marks the present organization of the system. The past therefore influences the future, because what could happen to an emergent system is constrained by what has already happened.

The emergent system we call a river illustrates how history is recorded by dynamic interactions and how an historical record influences the future. The channel of the river, its depth and meandering, can provide an accurate cumulative history of past flows of the river. Students of geography, geology and paleontology can picture past events by consulting old river beds. But the river channel not only records the river's past, it constrains the river's future. The channel is a robust attractor. The river constantly remodels its banks and deepens its channel, but a sudden change in the river channel requires a catastrophe, an earthquake or a volcanic eruption. Thus the ongoing evolution of a river is causally affected by its past history. So too are the future evolutionary channels of living systems chained to their past histories. Emerging species do not and cannot reinvent all their constituent parts anew. Creatures exploit intact, or in part, molecules and organs developed in ancestral species. Life is decked out in hand-me-downs.

Consider two more points when thinking about the histories of emergent systems like rivers and living creatures. First, such systems have histories because they are dynamic, they change over time as their interactions change. Change records time. Second, histories make individuality. The fundamental ingredients of rivers, for example, are uniform the world over; water molecules, the laws of gravity, and the laws of flow are the same for all rivers, past, present and future. From a reductionist viewpoint, all rivers should be essentially the same. Yet no two rivers are ever the same because every river has a unique history of interactions. The channels, banks, meanders, flows, and so forth have been, and will always be different. History makes individuality for rivers: for us too (§1, §2, §177).

§29 Emergent Adaptation

The second point about evolution is that it leads to adaptation. The word 'adaptation' is widely used whenever there is talk of evolution; but what does the word mean? Biological adaptation, at least to some minds, implies a sort of optimum arrangement. But it is not easy to define what is optimal about an arrangement. Adaptation may be defined in different ways; but adaptation, at the very least, means that a form of life has succeeded in surviving; what has changed persists. Adaptation, from this minimalistic point of view, need not reflect the best-of-all-possible worlds; adaptation merely means that a creature's interactions with its environment work well enough for the time being. The creature can extract enough energy and materials to preserve order locally (§7, §9) and maintain life and replicate, at least for now. The ecological niche, the system of interactions housing the creature, similar to the emergent river, is constrained by the limitations of its historical channel. There are certainly more efficient ways imaginable for producing salmon than having the poor fellows cross oceans to leap up waterfalls and die after spawning wasted millions of fingerlings; indeed, most of the salmon that now reach our platters have been raised in ocean pens. There must be cheaper ways to make people than sending them to college and buying them flats before they breed. Adaptation is making do with what past history allows the emerging system to modify.

§30 Survival of the Fitted

Attractors, like adaptations, describe dynamic interactions that tend to persist. An adaptation thus can be viewed as an attractor; the system resides in a particular state because that state works, that state 'attracts' the system. Defining adaptations as attractors gets us around the thorny problem of measuring 'improved' adaptation. Attractors can exist without necessarily being optimal solutions to problems of survival. Attractors persist merely because the system is built the way it is. Adaptation is thus a state rather than a goal. Adaptation is a form of order (§7, §9). Evolution can be viewed as a 'contrivance' that sometimes succeeds in transforming energy into new information (§72, §73).

Note that we have departed radically from the standard views of evolution and adaptation. Adaptation defined as an attractor abandons the presumption of improvement. Evolutionary transformation here holds no betterment. Evolution is driven, not by Darwin's *survival of the fittest*, what's best, but by the attractor's *survival of the fitted*, what works.

§31 Laws of Evolution

Because the influx of energy into the biosphere is random or chaotic, the outcome of evolution is unforeseeable. As we discussed above, ends cannot work as causes (§4, §5). Thus, we might propose three descriptive properties ('laws') of evolution:

1. Available sources of energy create opportunities for creatures that can exploit the energy,
2. Vacant space will be invaded and occupied by creatures, if it is at all possible for them to make a living in such space. Space here refers not only to physical room, but also to a more abstract room, that created by information (§7). A desert, for example, and a tropical rainforest may both occupy similar areas of physical space, but the rainforest provides more space for a diversity of creatures because of its rich organization of flora and fauna, the biological information that already exists there. Or the way a creature's body is organized, its content of information, creates space for parasites that otherwise may not be able to exist. Or more abstractly, a thriving economy, which is nothing but an organization, makes room for new businesses. So the space available for living is both physical and informational.
3. Nobody can know ahead of time where the process of evolution is heading, or what it will produce. In other words, energy and space combine to generate attractors for an essentially blind evolutionary process. In fact, we can say that the process of evolution is driven by a creature's eviction from its historical basin of attraction, and its suitability to participate in another attractor setup. However, the existence of a suitable basin of attraction does not guarantee that the candidate creature will be able to exploit the situation. The new attractor has to be vacant to accommodate the creature. There may be better adapted animals that can compete with the creature for the

attractor space. The mammals, for example, did not evolve to take over the world until the dinosaurs made space for them by becoming extinct. Competition encroaches on living room.

§32 Accumulating Complexity

The second law of evolution, quoted above (§31), states that creatures, over time, will evolve to occupy available space, the abstract space created by information as well as the physical space created by geography. In other words, the ordered arrangements of systems and their interactions provide material for evolution. Later on we shall describe the process of evolution in more precise terms, but for now we can say that information breeds information, interactions foster new interactions. This attribute of nature has an important consequence. It is true that we cannot predict how things will go (the third law of evolution) and, indeed, the history of life on earth has been visited by recurrent catastrophes and mass extinctions; nevertheless, the degree of organization of life will tend, of itself, to create living space for the evolution of new working arrangements. In other words, evolution (the creation of creatures in attractors) automatically generates opportunities for further evolution. This may sound obvious, but it is not trivial. It tells us two important things about life on earth (and also about the evolution of brains and other cognitive systems, as we shall see later on). First, in the absence of the catastrophes that are bound to occur from time to time, evolution will tend to accrue more information and become more and more complex as it proceeds. Secondly, in the time between catastrophes, the rate of evolution can appear to accelerate; more information increasingly generates more space for new information.

An example of the apparent increase of information through evolution can be seen when comparing the simplicity of single-cell bacteria with the complexity of animal cells and their arrangements into functional tissues. The evolution of human culture may provide a more transparent example. Witness the exponential growth of human technology from its origins in small roaming bands of hunters and gatherers (hundreds of thousands of years), through the transition of the agricultural revolution to the emergence of cities (thousands of years), to the industrial revolution (hundreds of years), and on to today's informational revolutions (decades, years and months). Quite simply, self-organization and complexity propagate themselves by their very nature, as we shall see (§72, §73). One word of warning: things may seem to be progressing or improving, but in reality they are only becoming more complex (§42). Also remember, please, the third law: we can't know where it's all heading (§31).

§33   Forces of Evolution

The basin of attraction, as we have said, includes the field of perturbations that still allows a system to return to a state of long-term stability (§22). A living creature may find itself forced out of its former basin of attraction by two types of accidents. Internal changes in the creature, genetic or physiological, may make the creature less fit to survive in its typical environment; the attractor for a lion without teeth is no longer the Serengeti, but the zoo. Alternatively, changes in the environment may render the environment hostile to the creature's historically normative interactions; the Serengeti without zebras is no longer an attractor for lions, irrespective of their dental state.

Once out of the historical basin of attraction, long-term survival depends on the availability of a new basin of attraction whose attractor suits the creature. For example, the agricultural revolution of the human species has led to the emergence of domestication as the only viable attractor for chickens, cows, sheep, and camels. None of these species have an option for life in the wild, which was their attractor for millions of years before the 10,000-year-old agricultural revolution in human culture. Hazarding a circular argument, we may say that *survival of the fitted* means that an individual biologic system that hits upon an attractor survives, at least while the system stays within the basin of attraction (§30). But we need not apologize for circularity; the game itself, self-perpetuation, is circular. Chickens have to have what it takes to lay fertile eggs to make chickens that have what it takes to . . . Or if you prefer closer to home, we have to have what it takes to fall in love with the right one to have children who have what it takes to . . . A biologic system that fails to hit upon a vacant attractor can, like an errant comet (§19), lose circularity and disappear off the map. Evolution is the emergence of such living systems that have happened to land in attractors. Using the terminology of the Darwinian triad (§28), *variation* is driven by energy that modifies the creature or its world, *selection* is the process that tests the creature's successful participation in an attractor, and *perpetuation*, the reward of the successful, is inherent in the way attractors work.

§34 Darwin's Finches

Darwin saw the laws of evolution at work on isolated oceanic islands. The development of bird life on the Galapagos archipelago, which rose from the sea about 1,000 km off the Pacific coast of South America, tells the tale. On some of the islands are land birds that probably descended from a few errant finches that chanced off course from the mainland some millions of years ago. By now the original finch population has given rise to some dozen different species, who make their various livings in very un-finch-like fashion. Rather than living on finch-sized seeds in typical finch fields, the new finch species include cactus birds, ground birds, mangrove birds, and even woodpeckers who use thorns, instead of regulation woodpecker bills, to dig out tree grubs.

Elementary evolution was at work. Random energy fluxes over the finch generations caused finch mutations, which, in most cases, were harmful to the mutated finches and these birds died. However, some rare mutant finches unexpectedly turned out to be able to exploit unfamiliar foodstuffs, normally inaccessible to standard finches. Such mutated finches were, in fact, adapted by chance to a new way of life; they had fallen into productive basins of attraction. The birds bearing such lucky mutations thrived and developed, over time, into new species. Thus, finch evolution was pushed by mutation-causing energy into strange forms, and some forms were selected to survive because they chanced to fit the untapped sources of food energy and information on the islands. Only because the islands were vacant of competing land birds, however, could the mutated finches radiate into their new living arrangements. A thorn-wielding finch would quickly have been put out of business had there been any professional woodpeckers around. In other words, mutations drove some finches out of the finch basin of attraction, but rare mutants could survive because there was space to accommodate them in new basins of attraction. These mutated finches could take part in new stable interactions. However, available ecological space will not become a home to creatures unless the creatures have been pushed out of other, more familiar basins of attraction, and, conversely, mutated creatures will not survive unless they chance to land on basins of attraction that have space and interactions to accommodate them (§31). Attractors and species co-evolve.

§35   Evolution and Finality

Survival, even if only a meager and costly attractor, has value for life. And values are ends. Didn't Newton annul causal ends? Does Aristotle's final cause (§4) make its comeback in biology?

No classical physicist would think, or much less say out loud, that gravity exists to hold us securely on the surface of the earth or that ice floats to keep the deep sea alive. The properties of matter are what they are, and that's that. Yet professors of immunology do not hesitate to declare that the immune system exists to prevent infectious disease, or that the immune system functions to maintain the distinction between self and not-self. Has the concept of Darwinian adaptation through natural selection opened the door for teleological causality in biology?

§36   Push-Pull

Decidedly not. The difference between Aristotle and Darwin is the difference between a pull and a push, and that's a world of difference.

Aristotle, like any rational thinker before him, saw all of existence as existence in nature, and he saw nature as contained in the infinite web of time. Although nature moved in recurrent seasons and other cycles, it was absurd to imagine that nature had a beginning or that it was going anywhere it hadn't already been before. There could be nothing new under the sun, so an ancient Hebrew concisely summarized Greek thought on the matter (Eccl. 1:2–18). If innovation is impossible, the perfection of nature could only be its harmony. Now, nature is harmonious when each thing performs its natural function in the overall mosaic of existence. Thus, it was only natural to Aristotle that each thing be drawn to its final end by order of harmony. The teleology of a thing, our knowledge of its final function, was tantamount to nature's ultimate knowledge of itself, the way things should be. In other words, Aristotle's final cause is the fate of all things to be *pulled* into place, willy-nilly. Teleology was inherent in the Greek world-view (§4).

In contrast to the pull felt by Aristotle and the Greeks, Darwin and the Western tradition see nature pushed blindly from behind. Evolution's push comes from a continuous stream of energy impinging on the world (§9). The energy affecting life enters the biosphere from the sun and cosmos beyond, and wells up from the earth's internal heat and radioactivity. The effects of this energy on life are felt on several scales, from the microscopic, random physical and chemical perturbations that produce DNA mutations, to local variations in the environment, to more global processes like climatic shifts and continental drifts, to major volcanic or interstellar catastrophes that cause mass extinction of life. These chaotic processes not only mutate life, they select fit creatures for survival and eliminate creatures who happen to fit less well. The living forms, the recipients of this largesse of energy, respond as best they can, within the possibilities inherited through past evolution. The river of life digs its new channel, constrained to some degree by the banks of its historical channel (§28). Although the word 'attractor' means 'pull' semantically, we apply the attractor concept to natural dynamic systems, which are pushed thoughtlessly rather than pulled teleologically.

§37 Abraham's Causality

The primacy in Western science of the *push* has emerged from the idea of *progress,* an idea which was absurd to Aristotle, as well as to anyone else who saw nature as all that is. Contrary to ancient experience, the concept of progress assumes that things are heading to new vistas, to places where they haven't been before. Novelty is not only possible, it is ordained. The idea of progress, the idea that time progresses in one direction, as a fleeting arrow rather than as a cycling wheel, is a revolution in human thought that first appeared along with a monotheistic concept of history.

Ancient natural philosophy the world over, Egypt and Greece and India and China and the Americas, saw the gods as the expression of the powers of a pervasive nature. The emergence of Monotheism was not merely the concentration of the gods into the One God. Monotheism took God (and man too) out of nature and put both nature and man into the hands of God. The agency of God reduced nature to a mere object. Nature was no longer divine: she could be tampered with. The ancient Hebrews developed the novel idea that nature, like humanity, has a history. Things and time have forward motion. True, there may be back-sliding, but there is no fated, cyclical return of the wheel. The Hebrew idea of progress spread to Europe with Christianity. Renaissance Europe secularized the concept of progress by trimming off the initiating Fall and the final Redemption, married the concept of progress to Greek logic, and Western science was born of that union. The irreversibility of history (time) is now embodied, scientifically, in the second law of thermodynamics (§9).

§38 Descent of Nature

How strange that today Western science has no need of God, despite the birth of science out of a seminal conception of God (§37). Nature first lost her divinity when God was located outside of nature. God now may have lost his standing in the world-view of Western science, but nature, nonetheless, is left diminished, an unsanctified object liable to human intervention. In the primal days before the monotheistic revolution, nature was sacred, and so nature was untouchable and progress was absurd. Now, nothing is sacred, nature is touchable, and progress is natural.

§39 Human Intervention

The biosphere is no longer a mindless recipient of energy. Human activities, a product of the evolution of human mind and culture, are now a major force in fashioning the earth. In the distant past, molecular oxygen, first produced by the metabolism of ancient living forms and now produced by plants, changed the atmosphere and modified the world. Now it is the human mind, a different product of evolution, that marks the world. The organization of information and energy is always taxed by increasing waste, increasing entropy (§9). Indeed, the present growth and activity of the human population have led, as we shall discuss below, to the emergence of new diseases that threaten human existence. Can our brains enlist the help of our immune systems to prevent and cure the new diseases (§183)? We'll leave that question for the end (§185).

§40 Shorthand Finality

So when immunologists say that the immune system exists to fight infection (§141) or when I wrote above that 'the immune system is the guardian of our chemical individuality' (§2), the words are a shorthand code for a causal push, and are not meant to imply a causal pull. We may use language that sounds teleological, but we really mean that the evolutionary history of the system is such that the system now does something wonderfully complex. An animal may seem to have been designed with forethought because the complexity and aptness of its structure and behavior inspire awe. But design is apparent only to hindsight, as explained by Darwin. Things that work give the impression that they must have been made according to plan; the more complex, the more carefully planned. But that's only because we humans make things according to plan. Our way of thinking and talking smacks of finality because language and mind are comfortable with intentions. The intentions, however, are ours, not nature's. Attractors happen naturally.

COMPARATIVE SUMMARY OF EVOLUTION

Our treatment of biologic causality and evolution does not fit the standard view of molecular genetics.

§41 DNA Master Mind

The points I have been trying to make about causality and evolution in biology can be highlighted by comparison to a more traditional view. In 1961, Ernst Myer published a paper entitled 'Cause and Effect in Biology' (*Science* 134:1501–6, 1961). Wrote Myer:

> On the one hand is the production and perfecting throughout the history of the animal and plant kingdoms of ever-new programs and of ever-improved DNA codes of information. On the other hand there is the testing of these programs and the decoding of these codes throughout the lifetime of each individual. There is a fundamental difference between, on the one hand, end-directed

behavioral activities or developmental processes of an individual or system, which are controlled by a program, and, on the other hand, the steady improvement of genetic codes. This genetic improvement is evolutionary adaptation controlled by natural selection.

Myer states two principles here:

1. DNA codes are constantly improving through natural selection.
2. The development and physiology of each individual are reducible, in essence, to the molecular processes by which the master DNA program is decoded.

Myer distinguishes between two different biological disciplines that deal with these two principles: functional biology and evolutionary biology.

The functional biologist deals with all aspects of the decoding of the programmed information contained in the DNA code of the fertilized zygote. The evolutionary biologist, on the other hand, is interested in the history of these codes of information and in the laws that control the changes of these codes from generation to generation. In other words, he is interested in the causes of these changes.

Myer's enthusiasm for DNA as the master program of life may be a reflection of the heady days of optimism that followed the discovery by Watson and Crick of the DNA double helix as the molecular reality of the gene. Another ingredient in Myer's thinking was the application by biologists of information theory and computer metaphors to the study of life. Thus, DNA was placed at the summit of a hierarchy of information that resulted in 'teleonomy'. In the words of Myer: 'It would seem to be useful to restrict the term *teleonomic* rigidly to systems operating on the basis of a program, a code of information.'

Aristotle's *teleology* was thus replaced by programmed *teleonomy* as a mechanical explanation for the end-seeking behavior of life and evolution. The ghost of teleology orchestrating and directing the symphony of nature was now divested of its mystery, and revealed to be, at least in biology, the genetic code acting as a program. DNA was the captain at the helm, the cybernaut.

The idea of DNA as the master programmer of life has been embraced by molecular biologists, and even by educated laymen. But there are two problems with this idea. The first problem is that no evidence, indeed not even a benchmark, exists that might be used to demonstrate 'ever-improved DNA codes'. In what way is the 'lowly' gut bacterium *E. coli* less adapted to its way of life than is the 'high' human being who supplies the bug both shelter and livelihood? The adaptive supremacy of man cannot be found in number: there are more *E. coli* organisms in any one of us than there have been of humans in all the world for all of time (§180). Neither can the adaptive supremacy of man be demonstrated by an enhanced capacity to survive the insults and ravages of the environment: unlike humans, *E. coli* is immortal and more biochemically adaptable than are humans. True, humans are more complicated than are *E. coli*, and that's probably why humans are more fragile. But is complexity improved adaptation? The point is moot.

Stephen Jay Gould, like Myer a professor of biology at Harvard, has invoked Darwin himself to prick the balloon of human hubris. Writes Gould (*Ever Since Darwin. Reflections in Natural History*, 1977):

> In a famous epigram Darwin reminded himself never to say 'higher' or 'lower' in describing the structure of organisms – for if an ameba is as well adapted to its environment as we are to ours, who is to say that we are higher creatures?

DNA, despite Myer's assertion, is not 'ever-improved'. In fact, neither is DNA 'ever-new'; DNA is only 'ever-modified' (§73).

The second blow to the supremacy of DNA as the program of life, is DNA's fall from independence. DNA, unlike proteins, can indeed serve as the template for its own replication. However, proteins are the machines that replicate DNA. Moreover, proteins interacting with DNA control which segments of DNA are to be activated as genes, and when. A sheep called Dolly was cloned using nuclear DNA obtained from a cell in Dolly's mother's udder. The nucleus of the udder cell was placed in an egg cell. In its natural environment in the udder cell, this DNA instructed the udder cell to make milk. Upon transfer to the egg cell, however, the DNA from the udder made a whole new sheep.

The egg cell proteins 'reprogrammed' the DNA of the udder cell. Who's telling who what to do? The nuclear DNA or the cell protein? Is the chicken nature's way to produce more eggs, or are eggs nature's way to produce more chickens? Richard Dawkins has proposed that both the chicken and its egg are mere vehicles for the propagation of selfish DNA (*The Selfish Gene*, 1976). But DNA has no viability outside of its attractor arrangement. Indeed, DNA itself can serve as a tool manipulated for other purposes; the immune system, as we shall see below, constructs some of its DNA, somatically and not for inheritance (§111).

The biological world-view expressed by Ernst Myer proposes that the processes of life can be reduced to the decoding of DNA and that evolution can be reduced to improvement of the DNA codes. On the contrary, as I have described here, living creatures are dynamic systems that emerge from the connected interactions of their component parts in concert with the environment. Chickens and eggs, DNA and proteins, are essential parts of dynamic systems; neither member of each pair comes first. Attractors evolve no slave *vehicles* merely to serve the replication of Dawkins' *selfish DNA*.

Rather than being the 'improvement of DNA', evolution is the creation and occupation of attractors. One attractor is not 'higher' than are others, and a species that arrives at its attractor is not better adapted than is a species that has landed in another basin of attraction.

This basic democracy of existence does not negate the fact that some attractors, some long-term interactions, are markedly more complex than are others, or that greater complexity can unfold from simpler building-blocks (§181). Evolution can thus be creative without being hierarchical. The very existence of attractors, stable working interactions, makes new informational space (new room) for more complex attractors to emerge (§32, §81). Attractors propagate new attractors: the emergence of cells made room for the emergence of multicellular organisms. Indeed, the emergence of rivers coupled with the emergence of humans created the evolutionary space for emergent river-civilizations: the Nile, the Indus, the Yellow and the Jordan, and then the Thames, the Seine, the Rhine and the Mississippi.

§42 Real Progress

One final word of caution. Just because Western science is motivated ideologically by the idea of progress (§37) and just because we may see increasing complexity (§32), we must not assume that nature herself truly progresses. If there be any progress, it will have to be of our making.

Chapter 3
On Cognition

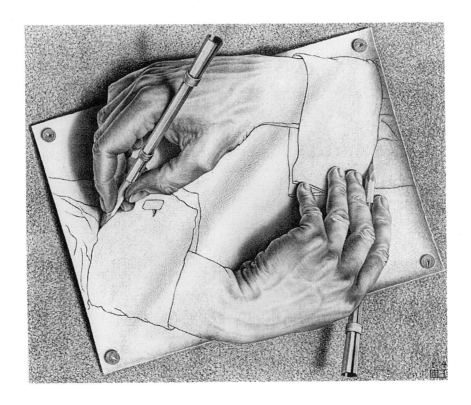

Chapter 3
On Cognition

DEFINING A COGNITIVE SYSTEM

To begin to consider the immune system as a cognitive system, we first have to define what we mean by a system and by cognition.

§43 Knowing

The word 'cognition' is derived from the Latin *cognoscere*, which means to know. To 'know' is an interesting word; knowing refers both to an internal state of awareness and to an objective test of performance. I am aware that I know, internally; I know that you know by what you do (or say), objectively. Both senses of knowing are nicely couched in the expression sometimes made in response to another person's surprising behavior: 'Does he know something we don't know?' Knowing in the biblical sense, too, contains a fine mix of internal awareness and objective action, with a twist of self-referential mutuality (Genesis 4:1). Adam and Eve know that they knew each other by what they did together; the reader gets to know too. The word 'recognition' (re-cognition) also preserves the sense of awareness–through–action inherent in cognition. Recognition is a recall of knowing, an awareness triggered by contact.

Now, if we insist that awareness is central to cognition, there is only one system, as far as we know, that knows with awareness: the human brain. Philosophers, psychologists, neurobiologists and computer scientists ponder how humans think with awareness (and how animals might do so), and ask whether computers could be programmed to think like humans. These experts explore in different ways the structure and behavior of the mind. The experts might disagree about what constitutes thinking with awareness, but all would agree that lymphocytes, singly or collectively, are probably never aware. So if we restrict cognition to *conscious* thinking, we can stop right here.

§44 Doing

But we may proceed to consider immune cognition if we relate the term
'cognition', not to thinking as a mechanism of awareness, but rather to
cognition as an operational strategy for dealing with the world. Cognition in this sense will refer to a certain way of adjusting to the
environment, of which the modality called human thinking is one
example. Our focus here will be on the instrumental aspect of cognition, on cognition as a game-plan and not on cognition as an experience.
From this viewpoint, we can ask whether the immune system, even
though it be thoughtlessly unaware, might carry out its duties using
operational principles similar to those used by the brain. So please put
aside, for a time, internal attributes of mind such as consciousness,
awareness, and intentions. The aims of the present chapter will be to
outline how cognition, purged of consciousness, might be defined functionally, and how cognitive systems might differ strategically from
other systems. The next chapter, 'On Immunity', will analyze the
particular components of the immune system and consider how they
generate cognitive behavior.

§45 Systems of Causality

The word 'system' refers to an arrangement of components that are
connected in such a way that they appear to some interested party to
form a coherent whole. I say *appear to form a whole* because one can
carve up the continuum of reality and define a system arbitrarily
according to one's taste, or functionally according to one's uses. Indeed,
carving up reality, as we shall see, will turn out to be a principal task
of cognitive systems (§57, §62).

In considering cognitive systems, or any type of system, it can be helpful to recall the paradigm of causality we discussed above in the section
on Reductions of Science (§6). For our purposes, let us define systems
as arrangements that organize causality. Units of causality, energy or
information (§7, §9, §10), enter the system from the environment as an
input, and the system, in turn, transforms and exports energy or information as an output. Hence, any system can be diagrammed as shown
in Figure 1.

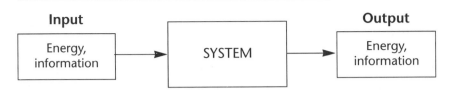

Figure 1: Diagram of a system

A flushing toilet can be viewed as a system that transforms the potential energy of the water in the reservoir into the kinetic energy of the flush. A radio receiver, like the one I'm hearing now, is a system for transforming an input of organized electromagnetic waves into organized sound waves. My ear, in turn, is a system that transforms an input of organized soundwave energy into an output of organized ion fluxes, the action potentials of the axons directed to the brain. (What the brain does to transform an input of physical energy into mental information is beyond present comprehension.)

Systems can operate at scales small and large (§15, §23, §28). The biosphere, for example, can be viewed as a worldwide system that transforms environmental energy, some of which is random, into orderly information, much of which is expressed in and by living creatures.

The solar system, whose sun supplies energy to the biosphere, is an arrangement of matter resulting from gravitational forces. Curiously, the solar system may not be a system in the restricted sense of a system as an *organization that arranges causality*; there is no intervening organization that mediates between the input of gravitational energy and the output of planetary movement. The orderly arrangements and pathways of the planets are a direct expression of gravity itself.

§46 Schematic Notation

Figure 1 illustrates a general format for depicting interactions. In this format, boxes designate entities, ideas, processes, or any object that we wish to mark as a definable unit. The arrows designate connections between boxes. An arrow exiting a box means that the item in the box produces, influences, or modifies the item at which the arrow is pointing. The boxes and arrows are to be taken as a convenience, as a way to picture relationships or interactions between entities. The Figure is a point of view, therefore, and not a fact. Ignore or modify the Figures, as you wish.

§47 Living Organism

To help us get the feel of systems as causal machines and to ease our path to a definition of cognitive systems, let us tarry for a moment to consider a living organism as a minimal system. What might be special about the inputs and outputs of a living creature that would set it apart from other, non-living systems?

The living organism can be viewed as a single system with two inputs and two outputs schematically drawn as in Figure 2.

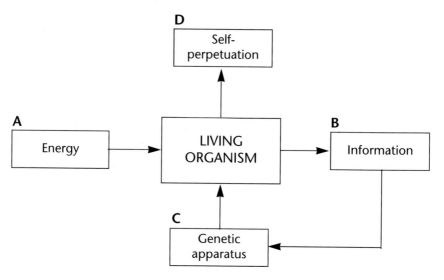

Figure 2: Diagram of a living organism

One input is *energy* (A) from the environment. This input includes food, light, heat, sound, pressure, electricity, radioactivity and whatever else hits or enters the body. (Some of these inputs may have a lot of information too, but let us ignore that for now.) The organism takes this flow of *energy* (A) and transforms it into *information* (B). We have defined information earlier as a form of order (§7, §8, §10); *information* here refers to all the organized molecules, physical structures, chemical reactions and behaviors that the organism produces from the input of environmental *energy* (§9, §31, §32). (Living organisms also put out a lot of energy, but we'll ignore that here too.)

Many kinds of systems transform energy into information: oceans, factories, economies, computer keyboards, or brains, for example. But a living organism has a special feature; included in its output of *information* (B) is a *genetic apparatus* (C). This apparatus is composed of genetic material (DNA) and the machinery for replicating, reading and expressing it (RNA and various proteins). Note that the *genetic apparatus* (C) is reflexive and self-referential; it emerges as an output of *information* from the living organism but it also loops back as an input into the living organism. The organism uses its DNA to build and regulate itself, as well as to procreate (§41). Thus, the *genetic* input/output (C) endows the living organism with *self-perpetuation* (D), an output that includes the capacity to survive and produce offspring.

In broad terms, we can call the A–B axis, the transformation of *energy* (A) into *information* (B), *metabolism*. The C–D axis can be called *reproduction*. Thus, a living organism is a system that *metabolizes* and *reproduces*. The living organism must metabolize because it takes energy and work to maintain order and information; so says the second law of thermodynamics (§9). The living organism must reproduce because, in time, it will succumb to entropy, disorganization. Thus, a living organism is a 'contrivance' for creating information, while paying its due to entropy.

Let me hasten to point out that there are problems with this definition. Consider *reproduction*. A particular living organism may not have actually reproduced itself, yet it still is to be classified as a living creature. Moreover, the organism may require a mate to reproduce offspring that emerge as a mixture of the parents' germ cells. Thus the full genome of neither of the parents is perpetuated in the child. Is this really self-perpetuation? The reader is invited to consider other paradoxes and problems, and perhaps he or she can more clearly define a living organism.

Problems notwithstanding, Figure 2 is useful for what it excludes. Viruses are not living organisms because they are a C–D axis (they are genetic material), but lack the A–B axis (they don't metabolize on their own). Figure 2 also excludes the earth, automobiles, and other machines that transform energy into information (the A–B axis), but do not reproduce themselves (the C–D axis).

I suppose we would be uncomfortable with our definition of a living organism if we ever succeeded in building a machine that could, on its own, extract energy from the environment and reproduce more machines. It is likely that Frankenstein's monster, a type of Golem, could metabolize – witness its energetic activity. However, there is no evidence the monster could reproduce, even potentially. Thus we still await the human creation of a living organism.

§48 System of Evolution

Above, we discussed evolution as a generator of attractors, stable interactions of creatures with their environments (§30). But evolution can also be viewed as a system. A system describing biological evolution is diagrammed in Figure 3.

Note that we have only changed some of the labels of Figure 2, a living organism, to make Figure 3, a system of evolution. Evolution, like

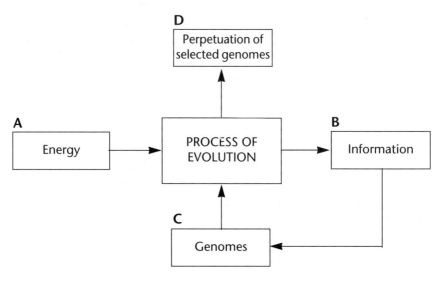

Figure 3: Diagram of the evolutionary process

life, transforms *energy* into *information* (§7, §9), some of which loops back to the system. The process of evolution is founded on the impact of environmental *energy* (A) on the creation and survival of particular collections of genes, *genomes* (C). Some *genomes* will perpetuate themselves in certain environments more successfully than do others (§31, §32). Thus, evolution expresses the effects of the A–B axis on the C–D axis. The reader is invited to develop this idea further. At this point, we shall return to cognition.

§49 Cognitive Systems

Living organisms are dependent on the co-habitation in one body of several different physiological systems, each responsible for performing some useful activity: nervous, musculoskeletal, respiratory, renal, intestinal, circulatory, immune, and so forth (§23). The inputs and outputs of the various biologic systems vary with what the system is designed to see and to do: a tap on a tendon (input) can lead to the jerking of a knee (output), a litre of beer (input) can lead to a litre of urine (output), a comely face (input) can lead to a marriage (output). What connects these inputs and outputs are systems which, as we have said, are arrangements for causally transforming energy and information (§6, §7, §9, §45). The first example (the knee jerk) and the third example (the marriage) are both transactions mediated by the nervous system, but only the decision to marry could be called cognitive. The knee jerk is a thoughtless reflex. In fact, the knee jerk has to be thoughtless; it won't jerk if we think too much about it. The second example (the production of urine) is also not cognitive, although it is sophisticated, complex and precise. The kidneys, in fact, regulate the volume and the composition of the blood better than can any mechanical device yet designed by the collective engineering talent of the human brain. Indeed, the production of urine by the kidneys is probably more precise and better regulated than is the process of courtship and marriage supposedly managed by the brain (the production of urine, unfortunately, may even have more adaptive success).

So what are the defining attributes of cognitive systems, if we discount the fact that a system, like the kidneys, may exhibit complexity, precision and regulation to a degree that might rival that of some nervous systems? Cognitive systems, I propose, differ strategically from other systems in the way they combine three properties:

1. They can exercise options; *decisions*.
2. They contain within them images of their environments; *internal images*.
3. They use experience to build and update their internal structures and images; *self-organization*.

Why have I picked just these three attributes to define cognitive systems? In the coming paragraphs, we shall discuss each of these features in some detail using examples drawn from the way our brains operate. Obviously, we are not going to discuss brain mechanisms, which are largely unknown, but rather the observable behaviors that emerge from brain mechanisms. The discussion will draw on common sense and common experience, and not on expert information. (Any questions the reader might have about the mind, or about the logistics and hardware of the brain, should be referred to professional philosophers, neurobiologists and cognitive scientists.) At the end of our discourse, the reader will see that *choice*, *internal images* and *self-organization* reciprocate conceptually in a cognitive game-plan for survival (§82); these three attributes in concert make it possible for cognitive creatures to interact with the world in a way that supersedes the confines of evolutionary genetics. Cognition will turn out to be a form of meta-genetic adaptation. Cognition, as it proceeds, creates individuality (§177).

DECISIONS

Cognitive systems make decisions; what could be a decision in a world managed by unconscious and deterministic causality?

§50 Options

Cognitive systems are notable in being able to choose among options. Input choices continuously challenge the brain. The radio is still playing while I write these words (it may be playing when you read them). The conflict between attending to the music and attending to the book is resolved by my choosing which input is to be the subject of my interest and which is to be the background.

Choice of subject commonly confronts people at parties. We may direct our attention to the words of the person standing at our side, while the words bombarding us from the other sources in the room are demoted to the din of background noise. But we may cock our ears to catch a more interesting conversation across the room, ignoring the words originating at our side. In other words, the central nervous system can choose to focus on one input out of a set of input options. Cognitive systems also have a choice in their outputs; we can talk to whom we wish at the party, or leave. We can put down the book and turn up the radio, or vice versa.

The attribute of choice can be added diagrammatically to a system as shown in Figure 4.

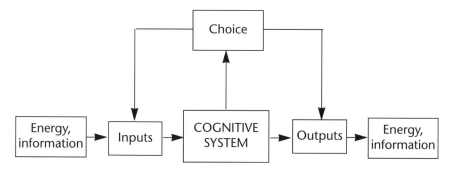

Figure 4: System choice

A cognitive system has two outputs (arrows out). One of the outputs, the *Choice* box, can determine (arrows exiting *Choice*) particular preferences among sets of potential inputs and outputs of the system. Note the circularity here: the system generates choices that affect the behavior of the system itself. We saw above that living organisms feature, in principle, circularity; remember the C–D loop that affects the A–B axis. Circularity, reflexive self-reference, is particularly vital to cognitive systems.

§51 No Options

All biologic systems collect input and dispense output, but not all of them can make choices. The healthy kidneys produce urine of a certain volume and a certain composition depending on the volume and composition of the blood flowing through them; they have no choice. The healthy lungs respirate without choice, the healthy heart pumps the required output of blood without choice. It is true that these biologic systems can deal with a wide range of inputs and respond with an appropriate range of outputs (§78). The healthy heart can achieve its needed output of blood, for example, by adjusting the frequency of beats, the pulse rate, to the volume of blood ejected with each beat. Physical training can lead to increased efficiency expressed as a lower pulse rate and a commensurably increased volume of blood ejected per beat. But such adjustments reflect regulation of quantities and not a choice among qualitatively different behaviors. The kidneys may be more sophisticated than the heart because they can adjust compositions as well as volumes. Both these systems, however, learn no lessons and exercise no options. The healthy heart or the healthy kidneys cannot be taught to refrain from pumping or excreting. To restate this point: cognitive systems are different from other systems in that they can decide among options.

§52 Choice and Determinism

What are decisions? We speak easily about options and choices because our brain has an awareness of choosing, and we generalize the feeling into a concept. But is it safe to rest a concept on a feeling? If the mind emerges from the brain, and if the brain is a material entity, then the brain must be subject to material causality. Hence, every action of the brain must have been determined by a preceding material cause, or set of causes. If this is so, then the mind must be deterministic and predictable. Where then is the power of choice? If we believe in choice, must we also believe in free will, and endow the brain with a mysterious ability to choose which causes it might wish to obey or ignore? If the brain is a deterministic system, how can it make choices?

The apparent contradiction between choice and determinism can be resolved by defining what we mean by choice. It seems to me that a choice truly exists, even in a deterministic system, if the system fulfils two conditions. The first condition is that the system, in its regular functioning, can relate to different inputs in a variety of qualitatively different ways. Upon hearing a tune, one's brain may instruct the body to dance, to sing, to sleep, or to exit the room, among many other healthy responses. In contrast, the non-cognitive kidneys can only make urine, the non-cognitive heart can only pump blood. Choices can be made only if alternatives exist. First there must be real options (§135).

§53 Internal History

The second condition for deterministic choice is more subtle. Although the course of action, the actual choice, is causally determined, the internal state of the system itself has a critical effect on the action taken. The person's past experiences, the person's mood and feelings, the person's intelligence, language skills, and so forth can all combine to affect the person's behavior. Indeed, no two people will behave in exactly the same way because brains are individualized by their history of individual experience (§1). The elements of individuality are so complex, even in their basic chemistry, that the individuality of the brain is unknowable in detail. Human behavior, therefore, is unpredictable in essence. True, we can guess how an associate is likely to choose, but we are often surprised. In fact, the degree to which we are surprised is a measure of the 'freedom of choice' exercised by our associate. Of course, the choice is still determined; it's only that we can never know all the causes, internal and external, operating on our associate. Note the loop; the system's history, its record of previous inputs and outputs, determines the system's present choice of inputs and outputs. Cognitive systems feature historical loops. In other words, cognitive systems learn from experience (§71, §75).

We can modify Figure 4 as shown in Figure 5.

These then are the ingredients of deterministic choice: the system has different options before it, and the unique history, the experience, of the system itself strongly influences the impact of external input on the system's behavior.

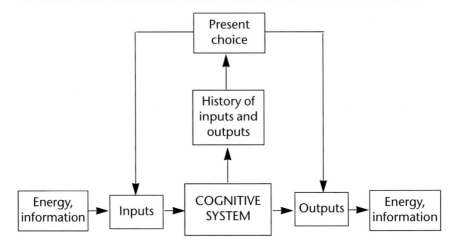

Figure 5: History influences choice

§54 Decision-making

The exercise of an option is a decision. What is a decision and how can a system make one? As thinking persons, we are aware of the decision-making process: we are conscious of a set of alternatives before us and we pick one or more for implementation, leaving the others to chance, oblivion or regret. Thus, cognitive decisions are easy to illustrate if we allow awareness.

But remember, we have ruled consciousness out of the game (§44). Are there unconscious decisions? Clearly there are. Have you ever apprehended a decision by hindsight? The moment the option no longer exists may be the moment you first grasped the fact that you had actually made a decision, an unthinking decision. Unthinking decisions are probably not rare; they often include, for good or bad, the most fateful choices of life: mates and professions. What then could be the mechanism of an unconscious decision?

§55 Particulars and Classes

The essence of decision-making, I want to suggest, is the association of a particular with a class. That sounds odd; let me explain. Decisions relate to particular instances and particular actions: I do this or I do that; I eat the meat or I eat the fish (I'm now writing this on an airplane); I take a taxi or ride the Underground (into London). The reader can supply his or her own, more or less momentous, decisions. In principle, decisions are particular acts that satisfy needs, and needs are classes of motivation (§65). The lunch trolley on the flight to London from which I must choose between the fish and the meat is a particular trolley. But the need to decide is a general motivation, a drive, which may have been triggered by the trolley, but yet is independent of the particular trolley. I choose the fish or the meat because I am hungry, or because, while not hungry, I don't want to be different from the other passengers, or I don't mind being different, but would like the approval of the flight attendant. Decisions have a common feature: classes of feeling have been applied to particular entities or actions. The entities (the fish, the stewardess, the taxi) are encountered in the external environment. The classes of feeling or need are inside us, they are internal. The decision emerges, therefore, from a match between an environmental case and an internal motive. Decisions are associations. Hence, we can say that cognitive systems make associations (§66).

§56 Finding Input

Decision-making is positive action; instead of passively receiving what the environment imposes, the cognitive system exerts its will (not free, of course, but still *its* will) in choosing among alternatives (§53). Cognitive systems are yet more resourceful: not only do they choose, they seek.

Seeking can be viewed as an extension of choosing. Boys and girls at the stage of choosing mates may not be satisfied by what's available at the moment, they actively seek the right one. We change channels to find the right program. We change jobs. We journey to the ends of the earth. The eye never rests, the hand is never still. We scan. We search.

§57 Reality Carving

The question here is obvious; what are we looking for, and how do we know when we have found it? The answer is simple in principle, but complex in fact. Cognitive systems 'know' what to seek because they are equipped with an image of their environment, built in. These internal images are critical; they filter the input, guide the search, and enable the decisions. Internal images carve reality into bite-size (§45, §62, §76, §172). How so?

COGNITIVE IMAGES

Cognitive systems create internal images of the world within which they make their living; what is an internal image and how does it work?

§58 Interactive Images

In a general sense, we can view parties to any interaction as images of each other. A physical example of such an image set can be seen in the interaction of a key and a lock (Figure 6).

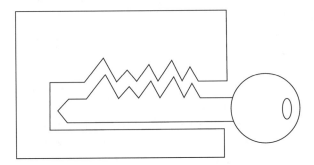

Figure 6: Lock-and-key image

The key fits the lock because the shapes of each are complementary in the right places. This steric fit makes each a negative image of the other; they reflect each other. This is obvious to the naked eye. With the mind's eye, however, one can see that the lock and key are not only concrete images, but also functional images of each other; the lock responds to the key by opening (or closing). This response, the interaction itself, creates an abstract image. Consider, for example, a skeleton key capable of opening 1,000 different locks. The skeleton key doesn't actually look like any of the specific keys whose function it has usurped. Yet, the skeleton key *works as if* it were a physical image of each of the 1,000 locks because it is a *functional* image of the 1,000 keys. So too do all interacting entities create functional images. Interactions arise through a mutuality of *information* between the interacting parties, irrespective of their physical shapes or material complementarity. Mutual images of interacting entities are encoded, abstractly, in their mutuality of information. Look again at Figure 6 and now see the complementing lines of the lock and the key as complementary fits, not in a physical space, but in an abstract interaction space. When we discussed the Laws of Evolution (§31), I pointed out that the space available to evolving creatures is not only physical space, but *informational* space. Informational space is room for interactions (§7, §29, §30, §32, §130).

§59 Interactive Matter

We have used the lock and key to illustrate an interactive image. In fact, if we were to look inside the lock and key on the atomic scale, we would see that the key turns the lock thanks to the internal electromagnetic interactions of matter that render metal solid and impenetrable. The physical fit of the lock and key at our world's macroscopic scale actually emerges from fundamental interactions at the atomic scale. Without such microscopic interactions, the lock and key would 'melt' into each other. Interactions are at the root of matter itself. What seems to be 'structure' at one scale may arise from what seems to be 'energy' at another scale (§6, §9, §15, §28). Scales establish the operational boundaries of interactions. Entities interact most naturally when they occupy similar scales.

§60 Adaptive Images

A creature, as we have also discussed, survives as an attractor, as a stable interaction with its surroundings (§19, §29, §30, §31). The successful creature is outfitted with the physiological systems it needs for this interaction. To extract oxygen from water, fish employ gills; to extract oxygen from air, land animals use lungs. Zebras have teeth suitable for eating grass. Lions have teeth suitable for eating zebras. The physiological systems of each creature fit the creature's lifestyle. Indeed, paleontologists can reconstruct the diets of extinct beasts by examining a single tooth: the image of the diet is encoded in the tooth. The sea is encoded in the fish's gills, the air in the lion's lungs and in the bird's wings. Physiological systems, effectively, are coded images of the needs they have evolved to satisfy. For example, the transition of animal life from the sea to the land is *represented* in the evolution of the lungs. Hence, structures *depict* functions. Recall our discussion of causality (§6). There we noted that structures are causally related to their functions. Thus if we know something about a structure, we may be able to conclude something about its function (and vice versa).

Of course, not all interactions rest on negative or mirror images. Some interactions push in the same direction; they are reinforcements, positive feedbacks. Husband and wife reinforce each other's mutual love, and their love for their children. Teachers and students reinforce their quest for knowledge. Corporations co-operate to make more money, and nations co-operate to make peace (or war). Note that positive reinforcements won't work unless they are co-ordinated, unless they are in agreement, in symmetry and in balance. Working together is an accommodation, a conformity. Common undertakings in families, friendships, schools, businesses, armies, governments or laboratories are possible through organization, hierarchies, rules, boundaries and limits. Any object with a boundary has, by definition, a form (§7). And any form with a counterpart is an image of the counterpart (and vice versa). Consider a glove: the inside of the glove is a negative image of the hand it fits; the outside of the glove is a positive image. The glove is also a more abstract image of the cold weather, or of the occupation or the style of the wearer (or of the culture that houses them).

§61 Images Serve

I have introduced image concepts because I would like the reader to consider the idea that cognitive machines like the brain and the immune system help the individual get through life by constructing internal images that map the environment. These internal maps tell the creature what to look for to satisfy its needs, tell the creature how to exploit the environment. Cognitive images can thus be seen as part of a game-plan for survival. Such images organize the interactions needed to maintain one's place in the flow of space and time. Interactive images serve to carve the intrinsically seamless continuum of reality into exploitable bits (§45, §57).

The internal images produced by the brain are a mystery; we know essentially nothing about their material reality. However, as we shall discuss later, the images of the environment encoded in the immune system have a chemical reality, made mostly of proteins. Some of these proteins are distributed around the body and form abstract, functional images, but many immune images are geometrical shapes (§130). But irrespective of how they are made, both brain and immune images are of two general types: innate images, which come with the genes we inherit from our parents, and acquired images, which come with the experience of life. We shall discuss acquired images below (§75, §84). The innate images include three types that are relevant to this discussion of cognition: feature detectors, attention preferences, and motive forces.

§62 Feature Detectors

Innate images operate, for example, at various levels of visual experience, starting from the retina, the light detector of the eye. The retina absorbs energy at the wavelengths of visible light and transduces this input of energy into an output of nervous signals legible to the brain. The brain, in turn, takes the retinal output as its input and fashions the pictures we see of the outside world. According to the disclaimer I made at the outset, we shall not consider the logistics of how the brain makes pictures out of ion fluxes, or how the mind enjoys the show; for that you will have to consult the professionals. We shall restrict ourselves to questions about strategy, about the game-plan for survival.

To enhance survival, the retina might be expected to act like one of those security video cameras that scan the visible environment and report everything to central control without bias. This expectation, however logical, is false; the output from the retina is not a faithful transcription of the photons of light that happen to bounce into the eye. The retina is 'hard-wired' to extract from the input of light particular types of information, and ignore other types. The array of light receptors is cross-connected by layers of cells in the retina in a way that enhances contrast, emphasizes lines, and detects movement. In short, the apparatus that collects photons is built with biases for certain aspects of the visible environment. The human retina is not a photographic plate, but a feature detector. It is built to report to the brain selected features of the photonic environment. The eye, as it were, sends the brain a fabricated version of the visual environment.

The brain (unconsciously) activates the eye muscles so that the eye continuously moves like a scanning device. The way our two eyes are situated allows each of the retinas to collect photons from the same objects at slightly different angles, an arrangement that provides the brain with input suitable for constructing three-dimensional pictures.

One may suppose that our predisposition to see moving silhouettes stereoscopically has served us well in the past as predators and serves us now as automobile drivers and tennis players. Indeed, the retina of each species is built to collect the data suitable to the way the species lives in its attractor (§19, §29, §30, §31): the eyes of bees are tuned to flowers, those of frogs to flies. The brain, in a sense, gets a sample of what it's looking for, what it needs to help the creature through life. Built-in predispositions, internal images such as these carve up reality according to specification (§45, §57); they encode information vital to the attractor arrangement that supports the creature.

A schematic diagram of selected visual input can be illustrated as in Figure 7.

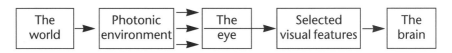

Figure 7: A cognitive eye selects visual input

The world, the external environment, is filled with energy that includes a *photonic environment* that bombards the eye with many arrows of input. The eye, however, is so constructed that only certain visual features of its potential input are transmitted to the brain, only a selected arrow is passed on through.

§63 Attention Preferences

The brain is born with basic images at 'higher' levels of integration too. For example, humans have innate preferences for looking at the human face. It has been observed in controlled experiments that human babies, within an hour or two of birth, are drawn to gaze on representations of human faces such as that in Figure 8.

Figure 8: A preferred face

Babies spend less time gazing at scrambled versions of the same visual elements, such as that in Figure 9.

Figure 9: A scrambled face

The preference of babies for looking at the human face was found to change over time in a complex way. For example, the baby's interest in the human face, present at birth, declines at one month of age, and then this interest returns at two months. But by five months, the developing child may actually be attracted more to weird faces than to the familiar, canonical human face. Curiosity, it seems, is also programmed into human development.

Attention preferences can be illustrated as shown in Figure 10.

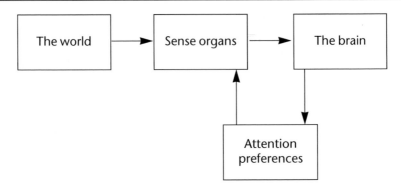

Figure 10: A cognitive brain directs attention

The brain directs the eye and the other *sense organs* to obtain particular types of information from the environment (*the world*) by way of built-in *attention preferences.* Attention preferences act as feedback loops through which a cognitive system can select its input, can see what it's looking for.

§64 Receivers Transmit

The dominant feature of the human environment is human society, an attractor supported by our lifelong fascination with all things human. Like the predisposition of our visual apparatus for faces, our auditory apparatus is predisposed to the sounds of human speech, our sense of touch to human skin. These predispositions are reinforced by the social group; our attentions themselves generate a feedback loop that directs the attentions of our fellows. The human eye, for example, both collects information and transmits information. A white ring of sclera surrounds the human pupil in a way that allows any human observer to see the exact direction of another human's gaze. The variable size of the pupil transmits information; the pupils can signal interest (dilated pupils) or disdain (contracted pupils). In times past, ladies used plant extracts (*belladonna* – 'pretty woman' in Italian) pharmacologically to dilate their pupils as a come-on. Indeed, human eyes are framed by eyebrows that enhance their visibility.

By their alignment, our brows declare our emotional response to what we see. Brows can be raised in wonder, contracted in concern, or knitted in censure. Brows may be plucked and redone to signal courtship interests. A universal signal of welcome among humans is a quick raising of the brows, a brow flash. This signal is usually transmitted unconsciously and received unconsciously, but it still works to declare a welcoming frame of mind. Look for the brow flash when you come home.

Think how much we depend on the information, mostly unconscious, we receive from the other's eyes. A brief mask of the eyes hides the entire face. A man who hides his eyes behind dark glasses may broadcast a threat; strangely, a woman in dark glasses broadcasts something else. Note that the dark glasses of a blind person present no problem; if the eyes do not function as receivers, we do not expect (unconsciously, of course) to see them as transmitters. Contrast the benignly dark glasses of the sightless with the hostility of mirrored glasses. The mirror conceals our neighbor's eyes behind our own reflected glances. We expect the eyes to broadcast intentions and not only attentions. Looking is communication.

Thus, the eyes, organs designed for receiving information, also serve as organs designed for transmitting information (§7, §8). But this two-way attribute of the eyes is not merely a poetic oddity; there is a principle of biologic signaling here, and we shall see it working at the molecular level in the immune system (§121). Reliable signals are the best signals; signals mean the most when they are guaranteed to be true, when they are fake-proof. It's hard to fake it when the receiver is also the transmitter.

§65 Motive Forces

The basic internal images created by feature detectors and attention preferences are complemented by primary motivations. Feature detectors and attention preferences regulate what enters the brain; motivations are classes of internal factors that influence the outputs of the brain: decisions, choices, actions. The internal motive forces of the brain are what we call emotions, feelings or affect (§55).

We are born with broad categories of affect that include love, fear, anger, taste, distaste, gratification, pain, pleasure, hunger, thirst, curiosity, and others, all adjustable and blendable. Since these affects trigger particular behaviors and reflect states of mind, the affects constitute functional images of certain activities. Emotions, as the word implies, move us to act. They influence how we behave because we learn to associate emotions with particular persons or particular entities located in our world.

§66 Meaningful Associations

The association of emotion with objects or ideas generates meaning because emotions trigger action. Above, I defined meaning as the impact of information (§7, §8). A perception can be viewed as an input of information; we sense *something*. However, if there is no response to the perception, then we may say that the information is meaningless. But, if the information generates a response, even if the response is only a change in state of mind, then the information has an impact; it bears meaning. Meaning is what information does. Indeed, the combination of information with affect, which generates meaning, gives rise to behaviors that feed back to influence the cognitive creature's world (§29, §30, §31). From this point of view, we can see that meaning is a type of interaction of a creature with a world. Figure 11 illustrates the point.

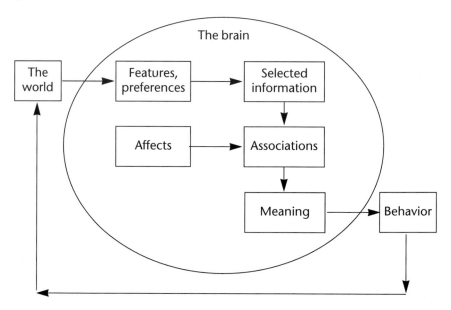

Figure 11: Cognitive interactions

The brain, through its *feature detectors* and *attention preferences*, collects *selected information* from the external *world*. *The brain* associates *affects* with the *selected information* to generate *meaning*, which is expressed as *behavior*. *Behavior*, what the cognitive creature does in response to an input of information, can feed back to modify the external environment, *the world*. A man who associates a feeling with a certain woman, may marry her to establish a dynasty. Another man may perceive the world's need for sturdy trousers, or for computer programs, and establish a bluejeans company, or a software company. Such enterprises have generated economic–industrial attractors that have influenced the behaviors of millions of other humans.

§67 Body Language

Important cognitive interactions, as we discussed, are social. We communicate our emotions to our fellows through physical expressions, the images of emotion broadcast by our behaviors. We have spoken of the eyes, but the face, the hands and the rest of the body too tell our fellows about our inner state. Body language is composed of innate public images: faces (and bodies too) can smile, frown, show shock, hate, love, peace, tension, interest, boredom, care, derision. Much of our body language is broadcast automatically and received automatically. Who can lie or love with a straight face? Children born blind, who have never seen another human, speak the body language of the sighted; they smile when happy, frown when unhappy, flash their eyebrows in greeting or raise them in wonder, shrug their shoulders in indecision, and stamp their feet in anger. Our inner emotions are encoded in body images legible to other humans, across cultures and around the world. Who can ignore the cry of a baby, a most compelling example of body language?

§68 Image Power

Learning is important to cognitive systems (§53). Let me then relate a lesson in image association. In my residency training in pediatrics, I had to confront the occupational problem of the pediatrician: how to examine an infant without triggering the awful cry. I, like all pediatricians, had to face two forces: Baby and Mother. Since babies are favorably attached to their mothers, it is reasonable to expect that

Mother is the doctor's safest connection to Baby. (Mother also is the one who makes the doctor appointments.) Therefore, young pediatricians (and many old ones too) make an effort to enlist Mother's confidence and support by assuming a face for Mother that broadcasts concern, responsibility and competence. After all, a satisfied mother is doctor's hope for getting a satisfied baby. So the doctor puts on a concerned, responsible and competent face like this for Mother:

Figure 12: The serious doctor face

Mother is heartened, but Baby cries. After much unpleasantness, I finally discovered a way to get around the problem. Instead of looking at Mother with the Figure 12 face, I learned to approach Baby with a face like this:

Figure 13: The baby-directed doctor face

As long as I maintained this face for Baby, I found I could auscultate the heart and lungs peacefully (provided I pre-warmed my hands and the stethoscope, to satisfy an innate tactile image), palpate a soft tummy, and even introduce a tongue depressor into the mouth without inducing spasm. The ears were problematic because facial contact with the patient had to be severed. But, whatever Mother may have thought at first of the doctor with the grin, she too usually ended the examination with her own grin. By then, the doctor's smile had become genuine.

This personal experience was not a controlled experiment, and I cannot report any statistical significance, or comment about the effects of the age, gender, genes, or past history of either the patient or the doctor. Nevertheless, the experience, for me, was highly repeatable, and it called my attention to the fact that humans arrive equipped with feature detectors, that certain features attract attention, and that subtle sensory inputs can be linked to emotional outputs of great power. Fortunately, you don't have to take my word for it; you probably know the power of a smile, or of a frown.

§69 Image Dysfunction

Internal images, like other physiological properties of systems, can fail to develop, or may function improperly. Image dysfunction may be trivial. For example, there are people who congenitally suffer from a poor sense of direction, get lost easily, and have difficulty locating where they parked their car. Here the defective image is an inadequate internal map of the physical environment. But image dysfunction may also be incapacitating. Autistic children, for example, appear to be unable to form emotional bonds to other humans, including their parents. Could it be that such children lack the innate images of emotions required to bond to other people (§55, §65)? Could some habitual criminals lack the innate images of human society that make socialization possible? These questions might be answered once we know how to investigate the neurological basis of brain images. At present, we can only observe how brain images *appear* to function. Autism seems like a disease of image connections, but how can we tell whether it really is? Below, when we discuss autoimmune diseases, we shall see an example of image dysfunction in the immune system (§160).

§70 Mind Your Images

Not all serious image dysfunctions are rare. Consider racial and ideological prejudices; note the destructive images of heaven and of hell that have infected humans with violence. Images need constant mending.

SELF-ORGANIZATION

Cognitive systems organize themselves as they evolve; what is self-organization and learning? Biologic evolution is the self-organization of species. Individual and cultural self-organization is somatic. Complexity is progressive.

§71 Learning and Memory

The hallmark of cognitive self-organization is the process we call *learning*; a cognitive system, through experience, acquires new capabilities and behaviors. Once learned, the new acquisitions are stored for future retrieval and use. This is what we call *memory*. Learning and memory are so much a part of the human cognitive experience that no more need be said about them. However, we do need to discuss self-organization as a general process, because even non-cognitive systems can self-organize. What then is the essence of self-organization? It is the progressive creation of information (§7).

§72 New Information

In essence, the acquisition of new information depends on unpredictability. The argument goes like this: Claude Shannon has taught us that an output of information generated by an input of information is communication, or perhaps the transformation of information from one form into another. Mere communication, however, does not, of itself, create new information. For example, the information in a DNA sequence of a gene can lead to the synthesis of a protein. The sequence of amino acids in the polypeptide chain of the protein, however, is already inherent in the sequence of nucleotides in the DNA; there is no net increase in the degree of 'order' in the protein compared to that of the parent DNA. The DNA, in effect, has programmed the protein.

Of course the protein may *function*, as an enzyme for example, in a way in which the DNA cannot. Hence, the transformation of DNA information into protein information has an impact; has, in fact, generated *meaning*. But new meaning is not the equivalent of new information (§8).

New information arising within a system, true self-organization, has to have been unprogrammed. If it has been programmed, then it is not new; the protein, for example, is already spelled out in the system's DNA. What is programmed is predictable. Therefore, self-organization is the generation of a new order out of some degree of unpredictability. It is the creation of information out of the flow of entropy (§9).

§73 Noise and Redundancy

A formal theory of self-organization has been developed by my colleague, Henri Atlan (see Henri Atlan and Irun. R. Cohen, 1998). Atlan has proposed that random 'noise', what I have termed here unpredictability, is one of the two conditions needed for a system to self-organize. Noise is inevitable, says the second law of thermodynamics (§9). Noise generates change in the established order. If the change results in a new order, the system acquires new information. (The change could also destroy the system entirely, blow it up; in that case all information, old and new, is lost.)

But self-organization, as I just said, involves the progressive addition of information. Not only do we want new information, we want more information. How can a system add to its store of information? The second condition for self-organization, according to Atlan, is that there be extra, or redundant, copies of the old information. Quite simply, if the old information were not redundant, it would be destroyed in the process of transforming it into new information. For example, the mutation of an existing gene into a new gene necessarily destroys the information present in the old gene. To generate a net *increase* in information, it is necessary to maintain the existing gene, even while mutating it to create the new gene. But how can you have your cake (create the new gene) and eat it too (mutate the old gene)? Redundancy solves the problem: you simply get by with two cakes, one to eat and the other to have. Genetic systems create extra (redundant) copies of genes by a process called gene duplication. The redundant copies of the original gene are available as a substrate for mutation, leaving intact a copy of the original gene to preserve the original information. The addition of the new information to the old information produces the increase in information we have defined as self-organization.

A schematic example of genetic-self organization is depicted in Figure 14. In Situation (1), we see a background of information: a specific sequence of DNA (DNA.x) encodes a specific protein (Protein.x). There is no new or added information here. If the protein does some-

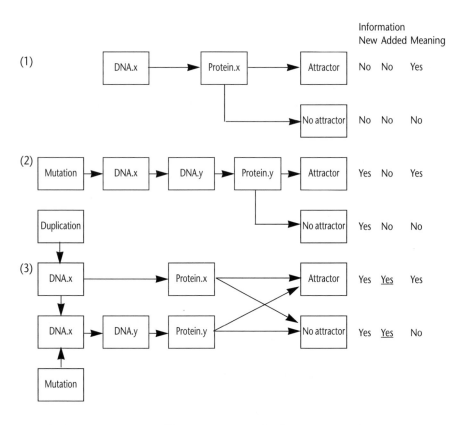

Figure 14: Genetic self-organization of information and meaning

thing, participates in an attractor, then the information bears meaning. If there is no functional protein (no attractor), there is no meaning. Situation (2) illustrates a change in DNA.x; a mutation has occurred that transforms DNA.x into a new gene, DNA.y, which encodes a new protein, Protein.y. Here we have new information, but no added information, no net increase in information. DNA.y and Protein.y have merely replaced DNA.x and Protein.x. Protein.y, of course, may or may not bear meaning. Self-organization takes place in Situation (3): A duplication of DNA.x makes it redundant, so a mutation of one of the copies of DNA.x produces DNA.y without the total destruction of DNA.x. Now we obtain two proteins, Protein.x and Protein.y, either or both of which could bear meaning.

§74 Self-organizing Brain

It is easy to illustrate the principles of unpredictability and redundancy in the process of genetic self-organization: mutations are unpredictable by definition, and duplicated genes are redundant by definition. The immune system, as we shall see later, also organizes itself employing unpredictability and redundancy at multiple scales (§129). But where is the unpredictability and redundancy in the cognitive self-organization of the brain? The unpredictability, obviously, comes straight from the environment; although we can predict trends and make reasonable guesses, input from the environment is always changing unpredictably (the weather, the economy, the government, our job, the needs and wants of those near and far). At the scale of life on earth, entropy rules and the world is fundamentally unpredictable (§9), and the brain has to deal with it. (At the scale of the Milky Way, our home galaxy where gravity alone rules, our earth is predictable, and negligible too. But how the brain minds that is another story.)

Redundancy is present from birth in the basic anatomy of the neural networks of our brains, and you may consult the experts for the details of how that redundancy is used by the brain to self-organize (see Gerald M. Edelman, *Neural Darwinism* [1987]). Here we shall make do with a behavioral example of redundancy: Affects, such as love, are essentially redundant because they can be associated with many different specific objects. Our ability to develop the complex networks of relationships that enable marriage, child-rearing, family life, friendships, social institutions, professional devotions, and patriotism involve an ability to associate various degrees of love to multiple objects. People who can associate love to only one entity, without redundancy, cannot organize their lives (although such situations have produced fascinating literature and theater; §11). We organize our lives through learning to navigate in the functional redundancy of our sea of emotion (§55, §65).

§75 Germ-line and Somatic Learning

Two types of learning take place in biologic systems: individual learning and species learning (§28, §53). Each of us lives by the grace of both. The evolution of the species can be viewed as a way the species, through genetic self-organization, learns to survive in a particular environment (§29, §30, §31). We receive the evolutionary endowment of our species through the germ cells of our parents, the sperm and egg that joined to make us. Because the genetic self-organization of the species is transmitted across the generations by the germ cells, species information can be called germ-line information. This germ-line information supplies us, for example, with the innate images of the human attractors we discussed above. Each individual begins life with the germ-line information he or she has obtained through the union of his or her parents, but then the individual goes on to create more specific and more diversified information gleaned from the experience of his or her own private life. This individual learning can be called somatic self-organization. *Soma* in Greek means body, and the term somatic refers to one's bodily individuality.

Individual somatic learning extends the evolutionary germ-line learning of the species. Beyond the innate germ-line disposition of the human species to acquire language, each of us English speakers needs to experience English, for example, to get our brains organized into an English mode, rather than into a Russian or Greek mode. Beyond our innate germ-line fascination with the general plan of the human face (§63), each needs to experience real faces to learn the fine distinctions that characterize the faces of real people. We need to experience emotions to learn how to develop emotional ties: and so forth and so on. Indeed, is there anything we don't have to learn? Somatic self-organization is the differentiation and extension of innate germ-line images through the assimilation of new information about the world into the organization of the individual. Cognitive systems learn from individual experience.

§76 Guided Experience

If the individual's brain can organize itself out of the stuff of experience, why do we need innate images to begin life with? Why not start with a clean slate, a '*tabula rasa*', as proposed by the philosopher John Locke? Why not let somatic self-organization start from scratch?

Despite the high hopes of social reformers, it seems that the human brain cannot divest itself of its ancestors and start afresh. This is because experience can only be experience *of* something. Perception, the initiator of experience, is possible only if the perceived entity can be singled out from the background of inputs bombarding the senses. In other words, the only way to experience a thing is to have 'in mind' a category, an idea, to which to relate the perception. Unless you have a notion of what you are looking for, you won't know it when you see it. Reality in itself is undivided; to observe, the observer needs preformed categories that divide reality into usable portions (§57). The arrays of photons, for example, that bounce into the eye from the visual environment do not arrive as organized collectives that in themselves segregate into defined objects. Photons are photons; it is the brain that sorts the photons into pictures of the objects we 'know'. Experience is formed by the transformation of perceptions into 'objects' or 'ideas'. The music does not enter my ear packaged separately from the noise of the street. There is no way to extract subjects from backgrounds without some image, even a coarse image, of the subject to begin with (§58). Quite simply, it takes information to catch information (§7, §9).

Evolutionary images, therefore, are the basis for the individual learning process (§1, §53). The brain has to have a set of basic tools with which to carve up reality into categories before any experience can be registered. Without attention preferences and innate images, we could experience nothing; all would be a confusion of 'noise'. Innate images serve as scaffolds for building individuals. Look at it this way: experiences are interactions, and a blank slate cannot interact.

Self-organization needs priming, and, like it or not, we cannot do without the dispositions of our species (§29, §30, §31).

§77 Aristotle's Idea

Although he lived about 2,000 years before John Locke, Aristotle already knew that it was impossible to create a mind from a 'blank slate'. Recall Aristotle's classification of causality into four causes (§4). An entity, taught Aristotle, has a *material* cause (the nature of its basic matter), an *efficient* cause (the force that brings it into existence), a *final* cause (its purpose, its teleology), and also a *formal* cause, the idea of the thing, its category of being. As we discussed above, science prefers to attribute an object's existence to material properties that are independent of our ability to know the object or to use it (§5). The formal teaching of Aristotle (and similar ideas of other philosophers too) is outside the scope of our subject matter, but Aristotle's idea of form illustrates the long history of the understanding that perception, knowledge and individual experience all rest on a platform of pre-formed images (§60). There simply is no way to extract signal from noise without them.

§78 Germ-line Development

Developmental programs such as growth, sexual maturation, parenthood, graying hair, menopause, ageing and death are in our germ-line genes to a great extent. The progression of these developmental programs in one lifetime is not to be confused with true self-organization. Particular expressions of germ-line genetic information may become manifest over time, but the information has been there from birth.

Organs not only need to develop, they need to be made active. Muscles not used degenerate. Hearts and lungs not exercised become feeble. Bones not stressed lose substance and fracture. To preserve their functions, biologic systems need to be active. But use, like development, does not depend on new information. Hearts, lungs, kidneys, muscles, guts, blood, skin and other organs can respond to use, abuse or deprivation. They cannot, however, learn anything new from the experience. They do what they are programmed to do, no more.

Thus, there is a fundamental difference between somatic self-organization and the growth and operation of the body. The cognitive brain, through experience, adds individual information to species information. The non-cognitive systems can only develop and use the germ-line information they were born with. But let not the neurons scorn the germ cells. As Sigmund Freud has noted, sexual energy sparks cognitive ideas, and not only spreads genetic traits. Our genetic motivations make possible our meta-genetic experience (§55).

§79 Human Culture

Note that the self-organization of the individual (§1, §53) is the basis of human culture, the self-organization of societies. Cultures create information when men and women transmit their individual experiences using language. Culture is the cumulative experience of individuals transmitted horizontally (from person to person over space) and longitudinally (from generation to generation over time). Culture, like life, is a 'contrivance' for developing and maintaining information in the face of entropy (§9).

Culture, like evolution, is a collective process, an attribute of populations. The individual, nevertheless, is the vehicle of both types of collective development. The genes of the successful person contribute to the biologic evolution of the species (§48); the ideas of the successful person help advance the culture of the society. Thus, the self-organization of individuals is the basis for the self-organization of collectives.

§80 The News, Good and Bad

So we are stuck with living in two worlds: the germ-line, evolutionary world of our species and the world we create as individuals and collectives through cognitive self-organization (§75). Unfortunately, 'two worlds' is a metaphor. In reality, there is only one world, a world that shrinks in size as we humans continue to be fruitful and multiply. Some would claim that our genetic nature is out of joint, our genes encode a disposition too aggressive for such a small planet. Perhaps it is inevitable that progressive organization in one sphere gives rise to mounting entropy, disorganization in other spheres (§9). Either way, we have no choice but to take seriously the self-organization of the brain. Every experience, to a greater or lesser degree, is formative. Kindness helps form kind brains, violence violent brains, goodness good brains. Individual experience is transmissible, giving rise to societies and cultures, good or bad. We know that the children of holocaust survivors, though born free, bear the mark of their parents. How many generations will continue to be poisoned by the legacy of slavery? It just may be a good policy to try and provide children with beneficial inputs. The self-organization of the brain is a self-fulfilling prophecy: good and bad are rewarded in kind (§42). The brain, like any resource, must be husbanded (§178).

§81 Progressive Complexity

Atlan's formulation of self-organization (§72, §73) explains how self-organizing systems may develop complexity at an increasing rate, a phenomenon we observed above in the evolution both of biologic systems and human culture (§32). The existence of 'old' information (old genes, for example) makes it possible for chance (mutations), to create 'new' information (new genes). Hence, the greater the amount of information already present in the system, the greater the chances for the generation of added new information. In other words, self-organization over time can lead to an increasing rate of the process of self-organization. Information generates information, and complexities generate more complexity.

Thus, the basic character of self-organization can account for the periods in which the process of self-organization within a system may accelerate. Self-organization can positively propel itself forward because information is both the substrate and the output of self-organization.

Lest we conclude that 'progress' is inevitable, note that self-organization is driven by noisy uncertainty; self-organizing systems can be fundamentally unpredictable. Moreover, even when new information does appear, whether or not the information persists depends on its meaning (§8), on its landing in any attractor arrangement (§31). Indeed, the entry into a system of new information can drive the system out of its customary attractor and into disaster. For example, a family may be destroyed, and not only built, by new relationships and education. A brain can be destroyed and not only built by experience. New cultural information imported into the Americas by Europeans demolished cultures of high attainment. Accumulating complexity can be good or bad. More is not necessarily better. History evolves, but progress is uncertain (§42).

THE COGNITIVE STRATEGY

Cognitive self-organization is a way of dealing with the world.

§82 Cognitive Attractors

I have used the terms cognitive *strategy* and cognitive *game-plan* to present cognition as an instrument, rather than as an end in itself (§44, §49, §61). Cognitive systems, like other biologic systems, provide a mechanism for interacting with the world; they contribute to the particular attractor that constitutes the creature's existence. A contribution to existence can, with reason, be called a *strategy* or a *game-plan* for survival. The word 'strategy' is derived from the Greek *strategia*, the general's art of leading an army, and so the art of planning and management in general. *Game-plan* comes from competitive games, and designates strategy in a more playful spirit. But let us not be misled into thinking that cognition implies designs or aims. Evolution, as we discussed above, proceeds without design or intent (§31). Because cognition appears to fit a way of life, a cognitive system may look like a strategic game-plan for survival, but only by hindsight. The cognitive system is the product of fortune, not of design. So beware, *strategy* and *game-plan* are metaphors, short-hand terms for noting that cognition fits a way of life. I do not intend to imply that General Evolution had a plan in mind when he or she deployed cognitive forces on the battlefield of survival.

§83 **The Cognitive Picture**

Let us review the three elements that, in their integration, make possible the cognitive strategy: *choices*, *images* and *self-organization* (§49).

Figure 15 diagrams the individual's cognitive interactions with the world. The individual impacts the world through his or her specific acts, what we have called *choices* (§50, §54). *Choices* emerge (§12) from the interactions of one's internal *images*, both innate and learned (§60, §61) with the forces of *self-organization* (§72, §73). The world drives individual self-organization through inputs of energy and information, and by the drain of energy and information (§7, §9, §31). This interaction is cognition without consciousness (§44).

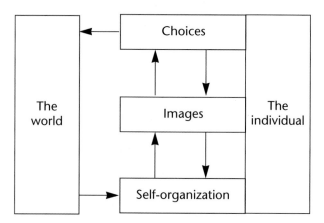

Figure 15: Elements of cognition

§84 **The Strategic Picture**

Individual cognition, like species evolution, organizes the interactions of creatures with the world; both processes participate in attractors. The question then is whether individual cognition differs in its survival strategy from the evolution of a species. Are cognitive attractors fundamentally different from the attractors of germ-line evolution? The answer, of course, depends on how you look at the question. At the scale of the big picture, attractors are attractors and survival is survival. Therefore, cognitive attractors and non-cognitive attractors accomplish the same 'ends', and so are essentially similar. When we look at the biosphere, however, we see that cognitive interactions add new levels of complexity to evolution; cognition enriches the diversity of existence.

Cognitive creatures, in contrast to non-cognitive creatures, learn individually and diversify as individuals, and not only as species (§53). Cognitive creatures establish homes and territories, generate families and tend offspring, migrate to new lands, and create hierarchical societies. In short, the attractor arrangements that include cognitive creatures such as the 'higher' vertebrates express more diversity, complexity, and opportunity than do the attractor arrangements of exclusively non-cognitive creatures such as bacteria or trees. Attractors that include learned behaviors produce cultures. Think of the increase in information that human culture, a brain-child of cognition, has brought into the world (§32). For better or worse, the world is not the same as it was before the emergence of humankind. The evolutionary appearance of cognition in living creatures has introduced a wave of new information into the world. Cognition is a seed-bed for novel attractors. Cognition has allowed individual experience to supplement speciation in world building. Cognition is yet another 'contrivance' for rescuing order out of the general flow of entropy (§9).

Figure 16 sums up the argument.

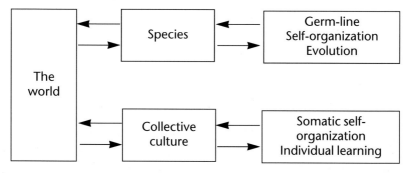

Figure 16: The whole picture

Cognitive creatures interact with their worlds as systems that self-organize in two different ways: genetically, by evolution of the germ-line; and somatically, by individual learning. The two forms of self-organization work in concert to enrich the biosphere.

COMPARATIVE SUMMARY OF COGNITION

How does the foregoing discussion of cognition compare with other approaches to the subject?

§85 Cognitive Reduction

The way I have presented cognition is not the way the professionals deal with the subject. Cognition is usually equated with the characteristic workings of the brain, as viewed from different aspects. At the basic nuts and bolts level, thousands of scientists are busy taking apart the nervous system to identify its basic building-blocks – the genes active in the brain, the protein molecules encoded by these genes, the way the molecules are regulated, and the cells that deploy these molecules. This is a reductionist program (§5, §6, §11). Many of the people engaged in analyzing the brain don't concern themselves with the emergence of cognition, or if they do wonder about the mystery of cognition, they tacitly assume that cognition will become clear once the last genes and molecules, the last nuts and bolts, are laid out on the table.

Many other scientists are involved in working out the wiring diagrams of the nervous system – which cells communicate with which other cells and how the cells are connected. At another level, the study of connectivity probes the functional effects of the connections – which connections activate cell signaling and which inhibit cell signaling. Cell connectivity has been studied traditionally by electrical stimulation and recording, and by the transport of labeled substances through cell fibers to other cells. Recent advances in optical imaging have made it possible to see in 'real time' the mutual activities of connected areas of the brain during the performance of brain functions – from simple movements and on up to intellectual activities, such as reading, speaking, or problem-solving. These connectivity studies constitute a transition from reduction to synthesis; they allow us to see how the component parts of the brain work in concert during particular brain functions. Connectivity establishes the physiological grounds that give rise to cognitive activities.

§86 Artificial Intelligence

The artificial intelligence (AI) people approach the brain from a different angle from that of the neuroscientists. Rather than looking at the hardware and wiring diagrams of the actual brain, they attempt to devise computers and computer programs that can carry out functions that are similar to those carried out by real brains. Implicit to AI is the hope that a given problem has one most reasonable solution. Therefore, if AI and evolution both take upon themselves the task of creating systems capable of learning, reading or recognizing faces, for example, then the 'right' solution devised by the computer scientists will be found to have been quite similar to the solution worked out by evolution in its design of the brain.

The hang-up for the AI program has been the failure of AI to devise a machine that can actually carry out a brain function to the brain's level of performance or competence. In other words, the AI people have yet to make a computer that acts like a brain, so there is as yet no way of testing whether the evolution of the brain has arrived at a similar solution. Moreover, evolution, as we have discussed, discovers only attractors, arrangements that work for a time (§28–§31). Evolution does not supply the 'best' solution to anything. So how can we trust evolution to arrive at the same solution to a cognitive 'problem' as does the AI computer scientist? After all, the computer scientist knows what he or she wants; evolution only plays around blindly.

The AI program, nevertheless, has succeeded in constructing very sophisticated and very useful machines for doing all kinds of tasks, some done better than the brain can do. The AI program has also been important in stimulating cognitive science people to try and define just what it is that the brain does, so that we might be able to recognize a brain-like machine if we ever see one.

§87 Linguistics

A notable attribute of the human brain, perhaps its most telling virtuosity, is language. Language is one way the brain can report its internal cognitive processes to outside observers; provided, of course, that the observers understand the particular language. The use of language, in fact, may provide some insight into the mystery of consciousness; what is consciousness, if not a way of talking to oneself? Be that as it may, linguistics, the study of the basis of human language, can provide some insight, it is hoped, into the way the brain is organized to perform cognitive functions.

Language has served as a probe for the cognitive organization of the brain in at least two ways. Anatomically, scientists have been able to identify distinct areas of the brain, centers responsible for a person's ability to understand and use language in different ways. These language centers were first discovered through the affects on language competences of injury to discrete areas of the brain. Directed stimulation and, more recently, brain imaging have confirmed and extended our basic knowledge of brain centers, their connections and their activities.

A second contribution of language study has been the appreciation that language competence arises through interactions between a person's experience and his or her genetic endowment; humans have an innate capacity to acquire language, but need to experience an actual language to realize this potential (§75). The innate language predisposition may include a 'generative grammar', a structuring in the brain of the basic rules of grammar needed to construct meaningful sentences (see S. Pinker, *The Language Instinct* [1994]).

The reader may wonder why I did not include the capacity for language within my definition of cognition. There are two reasons. The easy answer is that cognitive behaviors may be exhibited by non-human creatures, who don't use language. Any dog or cat owner knows that his or her pet learns, makes choices, and uses detailed internal maps of the environment, all without language.

The more subtle reason for excluding language from the definition of cognition is that language is only a particular manifestation of a more fundamental talent: the capacity to form and use internal images. Language is a way of abstracting an image, in this case a verbal–symbolic image, of some entity. At a concrete and superficial level, the word 'house' is an abstraction of the house, or of a house; the word 'Jane' is an image of Jane; the word 'love' is an image of an act, an interaction, or an internal state of a system.

Language is also a particular expression of cognitive self-organization. The *germ-line* capacity to use a particular set of abstract internal images, which includes deep rules of grammar, is the basis for the *somatic* organization of the nervous system acquired by experience, one's actual language. Thus language can be seen as a specific instance of the general capacity of cognitive systems to self-organize their internal images. We have used this general capacity, and not the specific example of language to define cognitive behavior. Nevertheless, language is distinctly cognitive, and we shall discuss below whether or not the cognitive immune system may be said to deploy an immune language (§137).

§88 Cognitive Practice

Cognitive science also includes scientific practitioners. Psychologists, psychiatrists, neurologists and neurosurgeons study the cognitive behavior of people (and of animals) in their various ways. They test and manipulate brain functions in health and disease, make diagnoses, and apply therapies. These activities and their successes and failures provide insights into the anatomy and the workings of the brain. Important theories of brain organization and function are based on such works. Consider the impact of Freud.

§89 Cognitive Philosophy

The subject of brain cognition, at once transparently available and yet obscurely complex to any thinking person, has supplied much matter for philosophical speculation. Descartes asked how might the mind, a manifestly intellectual or spiritual entity, be connected to the body, a material entity. Indeed, a significant portion of the history of philosophical discourse has been devoted to the problem of understanding, or bridging, the chasm between mind and matter. But there is no chasm; what we call mind is an emergent property of the material brain (§12–§18). Just as the property we call life emerges from the *interactions* of the material entities that make up the cell (§13, §47), so does mind emerge from the *interactions* of matter organized in the brain. Life is explainable by reference to the workings of the living cell, but life cannot be *reduced* to any particular set of material entities because life is an organizational entity created by their interactions. So too, mind is explainable as a creation of brain interactions; mind cannot be reduced to any discrete brain elements. Such is the nature of emergence. We can enjoy life, without having to posit vitalism, the existence of a 'vital principle'; we can use mind without having to posit the existence of a 'spiritual principle'. Emergence, from the aspect of Western science, is the *reduction* of phenomena to *interactions* of matter, rather than to isolated bits of matter.

§90 Carnal Knowledge

The approach to cognition outlined in this book obviously differs from the classical neurosciences; we have had very little to say about the component parts or wiring of the brain. We also have said nothing about AI mimicry of brain cognition, or about the views of cognition of psychologists, neurologists, or psychiatrists. We have not considered the views on cognition of philosophers.

The very different view of cognition presented here furnishes few points of contact with the cognitive concerns of the professionals, so there are few issues that are worthwhile comparing between such different world-views. The classical approaches to cognition all look at the substance of cognition, be it mind made of conscious intentions and language, or brain made of cells, molecules, and signals. Instead of analyzing the mind or the brain, we have defined cognition by the unique way in which cognitive creatures interact with their worlds. Rather than exploring substances, we have asked what may be unique about cognitive attractor arrangements. The answer has led us to define cognition as a function resulting from the self-organization of internal images, and from the behavioral options that arise from the historical self-organization of the individual. Cognition thus turns out to be a meta-genetic way for individuals to fashion unique arrangements, each with his or her own world (§84). Here we have removed consciousness from our definition of cognition, and have defined *meaning* as the impact of information, as an outcome of interaction (§8, §66).

In the beginning of this discourse on cognition, I referred to knowledge in the biblical sense as an example of cognition in which knowing refers to doing (§43). By this point in the discussion, it should have become clear to the reader that the biblical manner of knowing is not merely a literary sauce added to spice the fare; it is the meat of the matter. Adam's knowledge of Eve (Genesis 4:1) represents an interaction with consequences for the self-organization of the couple and of the child who emerges from that interaction. Meaningful interaction is the essence of cognition. In this sense, all knowledge is carnal. Our coming discussion of the immune system, I hope, will better clarify the point (§176, §177).

Chapter 4
On Immunity

Chapter 4
On Immunity

IMMUNE AGENTS: BASIC BRIEFING

We describe the cells and molecules comprising the system and briefly scan their general activities and their anatomical compartmentalization into functional immune districts. Our description uses a picture notation suited to illustrate the organization of immune agents into hierarchies and combinations.

§91 Immune Topics and Notation

The word 'immunity' comes from the Latin *immunitas*, which literally means *exempt from burdens of payment or taxation*. The word 'immunity' implies an untenable claim; even those of us who are immune to a particular disease are not exempt from continuous expenditures of energy and information to maintain health in this increasingly entropic universe (§9). What do we pay for immunity, and what do we get for it? For the answer, we need to look into the system itself.

Systems, like territories, can be visited in various ways, as suits one's objective or point of view (§45). Nevertheless, it matters how we arrange our route. Organizing the presentation goes a long way towards explaining the subject; the arrangement, as Shannon has taught us, is a message (§7). A full itinerary through a system would do well to attend to three aspects: the *agents* that do the job, their *arrangements in space*, and their *interactions in time*. Time, as we noted above, need not merely mark dynamic interactions; time can create history. Cognitive biologic systems, beyond their momentary physiology, can be characterized by their operations across two historical scales: the scale of the *germ-line* history of the species, and the scale of the *somatic* experience of the individual (§75, §84).

To suit the purpose of this book, our journey will skip the full itinerary through immune agents, space and time, and attend selectively to the highlights of immune cognition. We shall briefly mention the actual agents of immunity and the anatomy of the immune system. Readers interested in details of anatomy, cells and molecules may consult standard immunology textbooks or, better, read current review papers and surf the Internet.

Figure 17 presents a conceptual overview of the system.

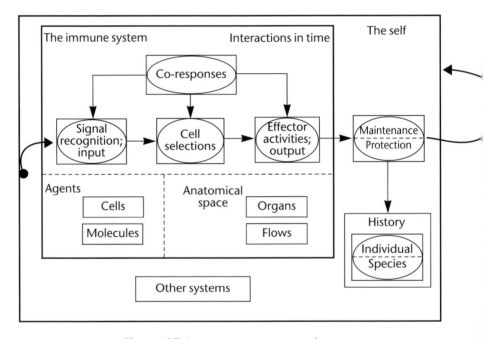

Figure 17: Immune system overview

The *immune system*, which along with *other systems* comprises *the self*, can be viewed from three aspects: *immune agents*, *anatomical space* and *interactions in time*. The system's agents are *cells* and *molecules*. Although cells and molecules occupy different physical scales (cells are actually packages of billions of molecules), it is useful to consider both as immune agents because some immune events are more clearly explained by the actions of cells, and other events by the actions of molecules. The system's anatomical space comprises organs and flows, and the agents of immunity spend their time in seeing (*signal recognition*), changing (*cell selections*), and doing (*effector activities*). The responses of immune system agents feed back as mutual *co-responses* to influence ongoing immune activities; immune agents adjust their activities to those of their fellows (§119). The system expresses both *individual* (somatic) and *species* (germ-line) history. Not only does the system *protect* the body against foreign invaders, its best known activity, but it attends to the body itself, acting as a *maintenance system*. This chapter will clarify this picture.

A word about notation: the schematic notation we used for Figures 1 through 16 has been a simple one (§46). From Figure 17 on, we shall use a more complex notation. The Figures now separate *objects* (designated by rectangles) from the *states of objects* (designated by circles or ellipses within rectangles). The advanced notation also allows the Figures to show *combinations* of elements; items separated by *broken lines* can be combined to generate joint products. For example, the broken lines in Figure 17 indicate that each of the interactions in time (signal recognition, cell selections, and effector activities) are performed by combinations of cells and molecules; and, furthermore, that all of these elements may be combined in particular anatomic compartments. Note an additional convention in Figure 17; the *immune system* box and the box labeled *other systems* both appear within the box labeled *the self*. This arrangement signifies that the immune and other systems are parts of the self. The new notation thus will allow us to show two principal properties of biologic systems: combinatorial modularity (broken lines) and hierarchy (boxes within boxes).

Arrows, as before, designate the directions of relationships or processes. The arrows in Figure 17 show connections between the interactions, and also illustrate that the signal input into the immune system originates from the self. The output of the immune system, a combination of maintenance and protection, feeds back into the self, which includes the immune and other systems. The immune system, as we shall discuss, is recursive; it expresses and records its own history.

I have adapted the format of boxes within boxes and broken lines from the precise visual formalism developed by my colleague David Harel (1987).

§92 Immune cells

Figure 18 names the major classes of immune system cells, also called white blood cells or leukocytes (*leuk-* means 'white' in Greek; when uncoagulated blood is separated by gravity sedimentation, the immune cells in the blood segregate as a white band visible on top of the more dense red blood cells; hence the name leukocytes).

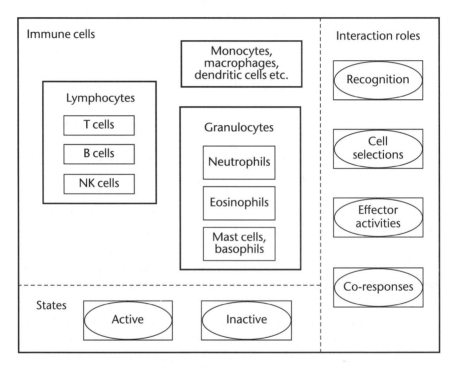

Figure 18: Immune cells

Figure 18 shows the immune cells divided onto three broad classes: lymphocytes, monocytes and granulocytes. Three kinds of lymphocytes are *T cells*, *B cells* and *NK cells*. *Monocytes* include *macrophages*, *dendritic cells* and a variety of cells resident in various tissues. *Granulocytes* include *neutrophils*, *eosinophils*, and *mast cells* (or *basophils*). Note the broken lines separating *immune cells*, *states* and *interaction roles*; the cells can, to various degrees, act in *recognition*, *cell selections*, *effector activities* or *co-respondence*. Each cell may also be in an *active state*, or in an *inactive state*. An active state means that the cell does something we can detect or measure; inactive cells are seen but silent.

Lymphocytes: the word 'lymphocyte' means cell (*-cyte*) of the lymph. The lymph refers to fluid that has exited the blood vessels and collected in the tissues. Lymph is drained from the tissue spaces back into the blood stream by a system of collecting pipes, the lymphatic vessels. *Lymph* in Latin means a clear fluid, and may be related to the Greek *nymph*, a water spirit. The name 'lymphocyte' is metaphorically apt because the clever lymphocytes are truly spirits of the body fluids. Lymphocytes include the *T cells* and *B cells* that can recognize antigens, molecular structures which other cells do not see. T cells and B cells are central to the cognitive enterprise because they can learn from experience. T and B cells can be said to be the princes of adaptive immunity.

B cells bear on their surfaces antibody molecules that function as sensors for antigens. After activation, B cells secrete their antibodies as free molecules into the surroundings. The free antibodies, like their mother B cells, can also recognize antigens, as we shall see later (§118).

T cells sense antigens using antigen receptors that are chemically distinct from those of B cells. Moreover, T cells can only recognize antigens that have been processed by other cells such as macrophages (§115). There are several distinct classes of T cells that, upon activation, can kill their target cells or produce signal molecules that activate or suppress the growth, movement or differentiation of other cells. T cells, unlike B cells, do not secrete antibodies or antibody-like molecules. T cells and B cells, therefore, endow the immune system with different agencies for recognizing antigens and responding to them. Why, you might ask, does the immune system need to sense antigens using two distinctly different cellular agents? We shall explore that question later (§119). But even at this early point on our route, we can see that the immune system can exercise different T-cell and B-cell options in sensing and responding (§50).

NK cells are lymphocytes that, unlike T and B cells, do not recognize specific antigens. NK cells, however, can recognize and kill cells that have been flagged by antibodies or that bear markers of abnormality (§117). Activated NK cells can also produce signal molecules like those of T cells.

Macrophages: if lymphocytes be princes, then macrophages are both kings and servants; macrophages, as we hinted, tell T cells what antigens to sense and, yet, macrophages carry out the molecular orders of lymphocytes. Macrophages circulating in the blood are called monocytes, but most macrophages seem to reside in particular tissues (skin, liver, eye, brain, gut and so forth) where, for historical reasons, they bear different names. Dendritic cells are a type of macrophage that is very efficient in activating T cells. The types of macrophages populating the various tissues help determine the type of immunological reactions that are characteristic of the particular tissue. One example will suffice for now; the nature of the immune response that can develop in the brain differs markedly from that which usually develops in the skin due, in part, to the particular types of tissue macrophages resident in the two organs.

Macrophages were the first immune cells to be noticed. Towards the end of the nineteenth century, Eli Metchnikoff observed with his microscope that mobile cells patrolled the body cavities of primitive creatures, and that such cells could ingest foreign materials and attack foreign intruders. Metchnikoff actually did his seminal experiment using rose thorns to agitate macrophages within starfish larva. The mobile cells were named phagocytes (*phagos* – to eat). Metchnikoff proposed that phagocytes were the primary agents of defense in 'higher' vertebrates as well as in 'primitive' invertebrates. The response of the individual to injury and infection was termed the inflammatory response, and macrophages (the *big eaters*) were envisioned to be the mediators of protective inflammation. The story of Metchnikoff and the beginnings of modern immunology can be read in *Metchnikoff and the Origins of Immunology. From Metaphor to Theory*, by A. I. Tauber and L. Chernyak (1991).

Metchnikoff's theory of protective inflammation by phagocytes was soon eclipsed by the discovery of the antibodies and of the lymphocytes that produced them. Antibodies could recognize specific antigens, while macrophages could not. For this reason, the macrophage was viewed as a 'primitive' immune agent fit for the likes of 'primitive' creatures. The macrophage was felt to have been superseded in evolution by the emergence of the smart lymphocytes, the agents of the truly adaptive immunity suitable for more 'advanced' creatures. Immunologists, intrigued by specific recognition, put aside the macrophage, and focused immunological research for most of the twentieth century on the lymphocytes, their antibodies, and their antigen receptors.

In recent years, however, the macrophage has made an impressive comeback. Immunologists discovered that T cells cannot recognize antigens without the assistance of macrophages (dendritic cells, in particular). Why should this be so? Would it not have been more efficient to have T cells that sensed antigens directly, without the need for help? We shall return to this question later (§119).

Granulocytes: these cells can be distinguished from macrophages morphologically by their smaller size. There are three classes of granulocytes that can be classified by their granules, small packages of chemicals located inside the cells: *neutrophils, eosinophils* and *basophils.*

Neutrophils are so called because their cellular granules were noted by early observers to stain a *neutral* color, compared to the other granulocytes: the eosinophils, which stain red, and the basophils, which stain blue. Neutrophils are not normally present in the tissues, but circulate in great numbers in the blood. They accumulate at sites of tissue damage, healing or infection. The numbers of neutrophils circulating in the blood rise quickly as the immune system responds to tissue damage or certain infections, and by counting the circulating white cells (the white cell count), the doctor can be aided in making a diagnosis.

Eosinophils are usually a minor fraction of the white blood cells (around 1%), but tend to increase in individuals infested with parasites. The functions of eosinophils are not well characterized.

Basophils in the blood probably represent the mobile phase of the cells resident in tissues called mast cells. These cells contain granules filled with histamine and other substances that produce the acute inflammatory responses associated with allergies. Mast cells can be activated to release their granule chemicals by certain types of antibodies.

§93 Immune Molecules

The classification of immune molecules is not simple. Academically, we might define immune molecules as the molecules studied by professional immunologists. Immunologists, however, enjoy academic freedom and are apt to follow their curiosity into neighboring or distant fields. Physiologically, we might designate immune molecules by their agency in immune reactions: an immune molecule would be any molecule that stimulates immune cells, or that is produced or activated by immune cells for immune purposes. I wrote 'for immune purposes' because immune cells, like all cells, carry out standard molecular housekeeping functions (§98), and housekeeping molecules should not count as immune molecules. But it's not always easy to draw the line.

Figure 19 shows one of many possible classifications of immune molecules and delineates seven types. *The antigen receptors of T cells and B*

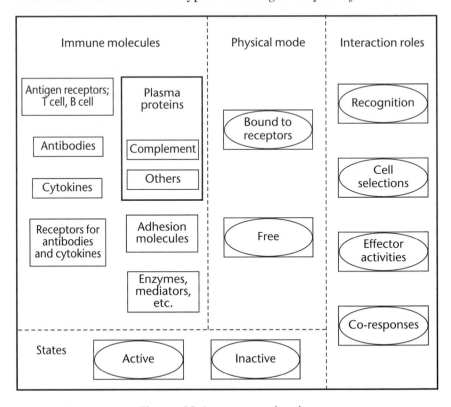

Figure 19: Immune molecules

cells and the *antibodies* are clearly immune molecules because they are exclusive to lymphocytes. *Cytokine* molecules are of central importance to immune phenomena; and some, like IFNγ (never mind the notation), seem to be produced only by immune cells. Many of the key cytokines (like IL-6) are produced by immune cells and by other cells too. *Receptors* for antibodies and cytokines are involved in the actions of immune cells, but are also found on many other types of cells. *Plasma proteins* like *complement* activate immune agents and are also activated by immune agents. *Adhesion molecules* are critical to immune reactions, but also to cells in general. Immune reactions, like other body processes, depend on certain *enzymes*, *mediators*, and other molecules.

It would be beyond the scope of this book to describe each class of molecule in the manner of a professional text. Here, we shall briefly characterize the various molecules.

1. *Antigen receptors* of T and B cells sense antigens and, upon doing so, rouse these immune cells into action.
2. *Antibodies*, produced by B cells, sense antigens and signal immune agents to act.
3. *Cytokines* (meaning to *activate cells*) are signal molecules that can activate the differentiation, growth, movement or death of many types of cells. Cytokines, which make immune activities meaningful, are produced by immune cells and by other body cells.
4. *Receptor* molecules for antibodies and cytokines are usually found on the surfaces of the responding cells. The receptor senses the antibody or cytokine and relays the signal into the cell, triggering the actual response.
5. *Plasma proteins* of certain types circulate in the blood, and can be activated by antibodies and other agents to produce immune effects. Complement molecules, for example, constitute a cascade of enzymes and adhesion molecules that can attract and activate phagocytes, or kill target cells. Such plasma-borne molecules are available for instant mobilization by immune reactions.
6. *Adhesion molecules* anchor cells to one another or to molecular surfaces. These molecules allow immune cells to stick to particular sites and to communicate with each other.
7. *Enzymes*, a class of molecules produced by all cells in the body, accelerate chemical reactions, either degrading molecules or building them. Among other uses, immune cells deploy certain enzymes

for penetrating into tissues and remodeling them, and for killing. *Mediators* are various molecules, usually relatively small, that react chemically to activate or inactivate target molecules or cells.

Note that immune molecules may reside in different physical states; they may be free in body fluids, or bound to other molecules of various types (receptors, for example). Like immune cells, immune molecules may occupy *active states* or *inactive states*, and the molecules, depending on their states and numbers, can play roles in *recognition, effector activities, cell selections* and *co-respondence*.

§94 Immune Anatomy

To complete this introduction, we shall schematically map the organization of the immune system in the *anatomical spaces* and *flows* of the body. The picture will show that the immune system is functionally and anatomically compartmentalized within the body; each tissue has its own particular needs for immune attention, and the various elements of the immune system are organized spatially to provide for these needs.

Figure 20 shows that *immune cells*, along with *other blood elements*, develop from *stem cells* in the *bone marrow*. Stem cells are metaphorically like the stem of a plant, the single basic element from which branch out all the specialized parts of the plant. Stem cells give rise to cells that *differentiate* into specialized cell types (analogous to the way the stem appears to give rise to the branches, leaves, flowers and fruit of the plant). But stem cells also must *renew* themselves, producing undifferentiated progeny to carry on the stem-cell line. The differentiated immune and other cells, which are born out of the pool of stem cells, enter the *circulatory system* and are distributed to the *peripheral lymphoid organs* (the *lymph nodes* and *spleen*) and to the other *tissues* of the body. (The lymph nodes are scattered about the body, and you have probably felt some enlarged nodes, at one time or another, in your neck or groin. The spleen is a fragile sack of blood cells located in the left side of the upper abdomen. The spleen, which can enlarge as a result of chronic immune stimulation, is susceptible to trauma and often has to be removed after automobile crashes because it doesn't stop bleeding.)

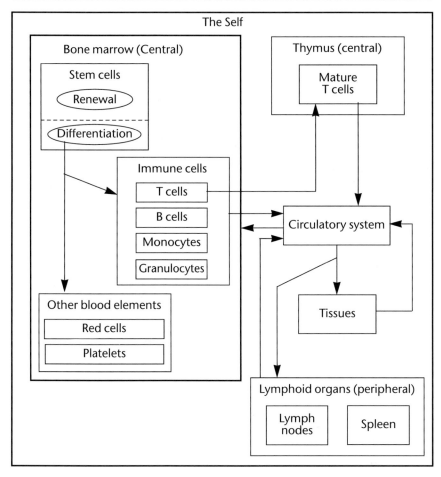

Figure 20: Anatomy of immune-cell development and flow

Most T cells have to sojourn in the *thymus* to *mature* before they can function. The thymus, where T cells mature; and the bone marrow, where immune cells are born, are called *central* immune organs. (The thymus is located above the heart, and the bone marrow, as you know from your soup menu, occupies the hollows of many bones.)

Figure 21 enlarges our view of the flows of immune cells. Immune cells journey between two compartments – the *tissues* and the *lymph nodes* (and other lymphoid organs) – using two sets of vessels – *blood vessels* and *lymph vessels*. The blood vessels operate as a high-pressure distribution system (the arteries) and as a low-pressure collection system (the

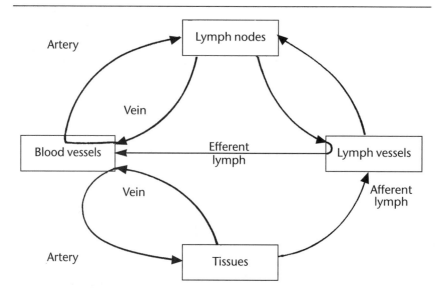

Figure 21: Blood and lymph flow connections

veins). The lymphatics have no central pump (heart), and operate as a low-pressure collection system only. Thus, the body contains one channel of distribution, the arteries, and two channels of collection, the veins and lymphatics. The *afferent* lymph vessels collect immune cells and lymph from the tissues and carry them to the lymph nodes, and the *efferent* lymph vessels carry on the lymph and immune cells to the blood for additional rounds of circulation.

Being heartless, lymph flows passively by gravity or by pressures exerted on the lymph vessels by contracting, moving muscles. Thus, unlike your blood flow, you can consciously control your lymph flow by controlling your muscles and by positioning your body. This control principle has practical consequences. Standing or sitting for long periods without moving, for example, can lead to swollen limbs, as your lymph stagnates. Stagnant lymph can also be beneficial; immobilizing a limb can slow the absorption of lymph-borne venom after a snake bite. Raising an injured limb above the level of the heart can hasten healing by creating a hydrostatic gradient that abrogates stagnation and so facilitates lymphatic drainage of the swelling. Some practitioners of 'alternative' health recommend standing briefly on the hands or the head at appointed times. The reader may supply other examples of moving lymph by effective positioning.

Figure 22 shows the body divided into three compartments – *lymphoid organs*, *interfaces* and *internal organs* – to make the point that different tissues require different immune services.

Each of the *interfaces* of the body with the outside world has special needs: the *skin* is variously abraded and exposed to radiation, heat and cold. Normally, the skin hosts certain 'harmless' bacteria and viruses, but the skin is also a gateway for pathogenic infectious agents, noxious

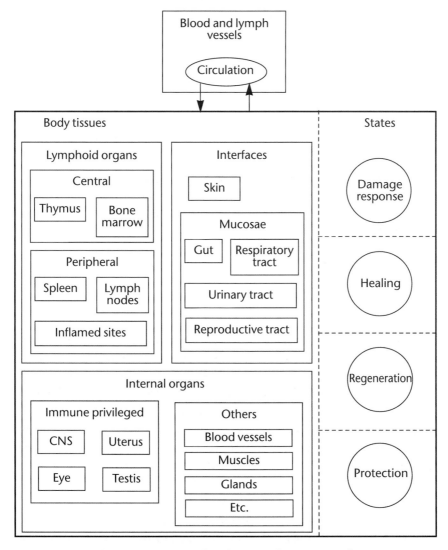

Figure 22: Organ distribution of immune cells

and foreign materials, and missiles from the environment. The *mucosal* surfaces (covered with a layer of *mucus – slime* in Latin) of the *gut,* the *respiratory*, *urinary* and *reproductive tracts* are open to the outside, and each has its special infectious and traumatic challenges. Foods and inhalants are foreign, but harmless; they require different handling at the mucosal surfaces than do infectious agents.

The other *internal organs* can be subdivided into those that appear to restrict the access or behavior of immune cells, the so-called sites of *immune privilege*. The central nervous system *(CNS)*, the *eye*, the *testis* and the *uterus* markedly limit the extent of immune interactions in their spaces compared to *other organs*, such as the *blood vessels*, *muscles*, *glands*, and so forth. The uterus, for example, accommodates the developing fetus despite its 'foreign-ness'. Fetal transplants to the mother's skin are rejected.

Immune cells are distributed in characteristic cohorts to each of the different organs, and the numbers and states of the immune cells in each reflect the varying local needs of the organ to respond to damage, to heal, to regenerate and to resist infection. For example, the liver can regenerate and the brain cannot; the gut will tolerate the normal bacterial flora and the blood vessels and liver will not; and so on and so forth. Each compartment has its characteristic immune needs and immune responses. The types of resident macrophages and the populations of lymphocytes that patrol the different organs determine the characteristic immune activities allowed there.

The *central* lymphoid organs (thymus and bone marrow) and *peripheral lymphoid organs* (spleen and lymph nodes) must dispose of abnormal, damaged and exhausted immune cells. The peripheral lymphoid tissues also filter pathogens and damaged body cells from the blood and lymph. *Chronically inflamed* sites in any organ will, in time, transform into *organized lymphoid tissues* with special local properties.

§95 Immune Districts

Now that we have introduced immune tissue compartments, we can take a second look at Figure 21. Consider that each lymph node drains lymph fluid containing immune cells and signal molecules from a particular region or tissue segment of the body. (The lymph may also carry cancer cells; so the lymph node may be the first colonization site of a spreading tumor.) Because each lymph node drains a particular tissue basin, the lymph nodes functionally segment the body into immune districts. The content of the afferent lymph provides each lymph node with immune information about the state of health (or disease) of the tissue region. The node can thus serve as a regional office for co-ordinating the immune affairs of its ward.

At the same time that it samples the lymph, the lymph node is visited and revisited by large numbers of immune agents circulating in the bloodstream. The lymph node is thus positioned at a junction between two channels of immune agents and information: a regional channel originating from the local tissue segment and a global channel circulating around the body as a whole. Using both local and systemic immune information, the node is in a strategic position to recruit resources from the blood to meet any local needs of its ward tissue. Thus the interactions of immune agents meeting in the node facilitate the reception and processing of immune information and make possible the recruitment of cellular and molecular reserve forces. These lymph node meetings are convenient sites for cognitive activities, as we shall see below.

Blood exiting the lymph node carries to the body and to the needy tissues the immune cells and molecules activated in the lymph node. You have all experienced signs of local immune committee meetings; recall the enlarged lymph nodes that appeared in your neck when you suffered from a sore throat, or in your groin when your foot was infected. That's your regional immune anatomy at work.

Chronically inflamed tissues will set up their own internal organized lymph node-like tissue to accommodate increased immune activity; immune anatomy is flexible.

Consider that the immune system, like the nervous system, must operate throughout the body; both systems care for the whole organism. Yet, each system uses a different topographical strategy. The nervous system, with rare exception, houses spatially fixed, non-renewable neurons that reach their target organs through a relay of variable but speedy electrical signals over fixed lines (axon and dendrite extensions of the neurons). The immune system, in contrast to the nervous system, is composed of constantly renewing, physically flowing populations of cells that patrol the body systemically in the blood and region-by-region in the lymph nodes. By selective adhesion, immune agents can exit their traffic conduits to accumulate at strategic sites when and where their actions are needed. Compared to the speed of nerve conduction, immune communication operates at the relatively slow pace of blood and lymph flow, cell migration and molecular diffusion. The nervous system operates a hard-wired network geometry like that of the telephone company (with one central exchange, the brain), while the immune system operates a mobile geometry with distributed stations (lymph nodes and spleen) similar to the strategy of the police, sanitation and fire departments and the board of health. A masterpiece of flowing anatomical organization creates a network that satisfies the varying immune needs of each body site. The subject of immune anatomy surely deserves more attention than we can give it here.

Figures 17 through 22 serve as a background briefing. Some of these pictures will take on added meaning as we get into the subject. The aim, as I said, is not to catalog facts about immune cells and molecules, but to show you the principles by which they interact to create the system. Let us begin from the end, by noting what the effector activities of the system can do for the body. What is the immune system really about?

MAINTENANCE

The immune system is famous for protecting the individual against foreign invaders, but it is just as active in providing maintenance. Immune maintenance of the functioning body, like the embryological building of the body, involves cell death as well as cell growth, movement, and support functions, depending, to variable degrees, on gene activation.

§96   Five Interventions

When we examine, without preconception, the actions of immune agents on the body, we can observe five types of effects. The cells and molecules of the system can:

1. make cells grow and replicate;
2. make cells die;
3. make cells move;
4. influence cell differentiation (which involves turning genes on and off);
5. modify tissue support and supply systems (which includes building connective tissue scaffolds, making cells sticky or unstuck, regulating blood vessel growth and blood supply, and disposing of waste).

Obviously, these five activities overlap to varying degrees. For example, the growth, death and movement of cells, and the development of blood vessels and connective tissues may result from the direct activation of certain genes by cytokines. Aspects of cell movement, mediated by adhesion molecules, and waste disposal, performed by phagocytes, may be less dependent on direct gene activation. In any case, how we elect to enumerate the list of immune activities is not really important. The important lesson is that the list of immune processes includes activities that take place before birth, during early development of the body. A single fertilized egg cell develops into the multicellular body of the mature organism as a consequence of these five actions. The fertilized egg becomes an embryo and the embryo becomes an individual person (or plant or animal) because particular cells replicate, die, move, and differentiate their special functions, all aided by a supporting and nourishing infrastructure. The immune system as such has little to do with these processes when they occur in the developing embryo before birth. Embryonic development is autonomously programmed and takes place without immune intervention. But after birth, the mature immune system is capable of implementing processes much more limited in scope, but nevertheless similar in principle to those that take place automatically during early development.

As any home-owner knows, maintaining a lived-in house amounts to building again (and again) different parts of the house. Clearly, the various structures of the house have to be redone at different intervals: paint, screens, windows, doors, masonry, siding, walls, roofing, plumbing and electricity require fixing or replacement at different times and to different degrees. The foundation, barring catastrophe, is not touched, while light bulbs need to be replaced frequently. In principle, maintenance of the body, like maintenance of a house, requires processes similar to those needed to build the structure at the outset. What was intrinsically programmed in the tissues during early development is later subject to decisions made by the immune system in response to the needs of the body. Actually, some of the cytokines deployed later in life by the immune system (IL-1, TNFα, TGFβ; never mind the designations; we'll explain them later) are also active during embryonic development. These cytokines are produced under different circumstances during embryonic development and during immune reactions, but the shared molecular signals demonstrate that the processes of development and immunity are related chemically. Using similar tactics, embryonic development builds the body, and immune reactivity maintains the body. Figure 23 illustrates this idea.

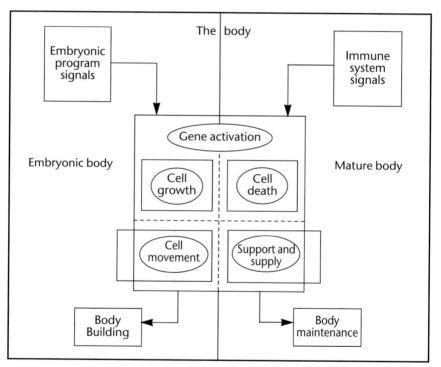

Figure 23: Embryonic development and immune maintenance

Five general cellular activities in combination (growth, death, movement, support and supply, and gene activation) take place in both the *embryonic* and the *mature body*. *Body building* is controlled by *embryonic program signals* and *body maintenance* by *immune system signals*. During embryonic development, the five activities are induced by chemical signals emanating from surrounding tissues; the body, as it were, pulls itself together. After embryonic development is completed, immune signals take over as the inducers of the five activities. Note that gene activation need not occur in all phases of cell movement or of support and supply; this is signified by the extension of these boxes beyond the borders of the gene activation box.

§97 Body Building and Cell Death

The details of early development are the meat of the science called embryology, and are beyond our present scope. However, one can intuitively appreciate that building a body from one fertilized egg cell requires that the cell grow and give birth to more cells. It is obvious too that cells must turn on the particular genes they need to allow them to differentiate into special tissue cells (brain, guts, muscles, blood, etc.). Differentiating cells must also turn off the genes that interfere with differentiation. It is also clear that the development of the multi-cellular organism depends on supporting tissues, blood supply, and waste disposal.

In contrast to these activities, the role of cell death in development is not intuitively obvious. The metaphor of the sculpture can serve to introduce the idea that destruction can be essential to creation: the differentiation of the block of marble into a work of art depends on how artfully the artist cuts, carves, chips and chisels: the statue emerges from artful destruction. So too is destruction required to build an organism. Fingers, for example, differentiate from embryonic limb buds by the programmed death of intervening tissue cells. The death of cells sculpts, as it were, a hand out of an embryonic paddle. Cell death is vital. How does the system kill?

The story of cell death begins with suicide; cells of multicellular organisms are equipped to kill themselves (some simple one-cell creatures can commit suicide too). This fact, of central importance to life, required over 100 years of scientific observation to come to the notice of biologists generally. The extension of Darwinism (§28) to cellular evolution might be to blame. The concept of the survival of the fittest seemed to imply that, at any scale, the fittest had to survive. Fit cells, like fit individuals, were likely to be those that managed to live, and not those that chose to die. Death, especially self-inflicted death, just had to be unfit. Despite this expectation, biologists have finally come to realize that cells are born with all the machinery they need to kill themselves, and readily use it. Cell suicide is natural. (Don't belittle the biologists for taking so long to notice the death program. Scientists, like other humans, tend to see best what they believe most; 'believing is seeing', we might paraphrase it.)

Programmed cell death has become a very important subject of study in embryology, immunology and cell biology generally. Cell suicide even has a Greek name *apoptosis* (in Greek, *apo* means *from* and *ptosis* means a *fall*). A visible example of apoptosis is the fall of leaves from deciduous trees in the autumn, which occurs because the leaf cells are programmed to kill themselves at the appointed time (the fall). All our cells, like leaf cells, carry genes expressing a cascade of signal molecules and enzymes whose activation leads to the suicide of the cell. But how can suicide promote fitness? Consider the molecule called p53.

One of the major contractors of death is p53. The p53 molecule is harmless as long as a cell is healthy. But let damaged cells beware; p53 is activated when it *senses* damaged DNA. The activated p53 molecule, in turn, activates various other cellular genes and molecules that can arrest the growth of the cell, repair the damaged DNA segment and, when necessary, trigger the cell's death by apoptotic suicide. The p53 molecule is known as a *tumor suppressor* molecule because it can trigger the suicide of cells with mutated DNA, abnormal cells that might otherwise grow to become tumors.

Since activated p53 can rid the body of potential tumor cells, most tumors have had to originate from cells with inoperative p53 genes. Thus, the development of a cancerous tumor requires at least two types of genetic (inherited) changes in the progenitor cell that begets the tumor: the incipient tumor cell must have a mutation that stimulates it, or at least allows it to grow progressively, and the cell must have a mutation that neutralizes the damaged cell's death program. Viruses, too, may take over host cells by inactivating p53 and other internal suicide molecules. Inactivation of the suicide system is clearly dangerous.

How then can the body rid itself of cells that, for the general good, ought to have committed suicide but have not? Fortunately, the immune system can step in and resuscitate death. Activated immune agents can kill the potentially dangerous cells that have lost their ability to commit suicide; the immune system functions as a fail-safe device to activate the internal death machinery of a renegade cell deranged by an infection or malignant mutation. Clearly the immune system has to make vital decisions; imagine the costs to health of sparing guilty cells or of killing innocent cells. Making such delicate decisions demands cognitive attention.

Immune death, once commissioned, can be executed by many agents: T cells by several different mechanisms can induce their target cells to perform apoptosis; NK cells can kill target cells by attaching to antibodies bound to the target-cell surface; macrophages can kill by phagocytosis and by producing chemical poisons; antibodies can circumvent the apoptotic mechanism and kill cells directly by activating complement enzymes that perforate the target cell membrane; and various cytokines (produced by T cells, NK cells, macrophages, or other cell types) can activate the apoptosis machinery of target cells.

Indeed, death is a two-way street; immune privileged organs (§94), like the central nervous system, may abort immune reactions by activating the apoptotic death of invading immune cells. Thus, the body is maintained by the regulation of cell death, no less than by the regulation of cell growth.

§98 Body Maintenance and Housekeeping

The immune system, broadly speaking, helps maintain the individual in the face of the blows, unpredictable but inevitable, that smite the body on its journey through life. The spontaneous accumulation of disorder, entropy, is effortless; it takes hard work to maintain operations (§9).

Note that the battle against the whims of entropy can separate maintenance from housekeeping. It is customary in molecular biology to distinguish between two principal types of genes: housekeeping genes and tissue-specific genes. Housekeeping genes are active in all cells at all times because the products of housekeeping genes are needed for the ongoing energy metabolism required of all living cells (§47). Tissue-specific genes, in contrast, are genes expressed only in certain cell types, when they are needed to carry out the specialized functions of the cell: muscle genes in muscle cells, hormone genes in endocrine cells, reproductive genes in germ cells, and so forth. But not all genes can be classified as housekeepers or specialists; maintenance genes are a third type.

Maintenance is like housekeeping in that it is required by all cells. However, maintenance contrasts with housekeeping in that housekeeping molecules are constantly needed, while maintenance molecules, such as p53, are needed especially at times of crisis, contingency or accident. (The distinction between housekeeping and maintenance is useful, but not absolute; maintenance molecules such as p53 can also perform some housekeeping functions.) Maintenance molecules, however, are not infallible. As we saw, p53 and other such molecules can be mutated into silence. The complex body requires the services of a maintenance system that can carry out its job with cognitive care and resourcefulness. The immune system is this maintenance system. Immunity takes the cell in hand when the cell's maintenance proteins falter.

Good maintenance requires, as we have outlined for the immune system (§91), a program of three parts:

1. Recognition – you have to see what is right and what is wrong.
2. Cognition – you have to interpret signs, evaluate results and make decisions.
3. Action – you have to actually do the job.

RECEPTORS

To protect and maintain the body, immune agents have to receive information. Biologic recognition involves both sensing signals and responding to them. Receptors have combining sites that sense ligands, and reaction sites that cause effects. The molecular specificity of receptors emerges from protein conformation (folding) brought about by non-covalent molecular forces.

§99 Specific Signal Recognition

We devoted space early on to discuss information and energy in some detail (§6–§9) because information and energy are the substrates of life and evolution (§31, §47, §48). We are now going to discuss *specific recognition* in some detail because the reception of specific molecular signals is the substrate of biologic interaction. Signal recognition, like all things biological, involves information and energy.

Signals create interactions. The world of elements with which you can interact defines the world in which you live. Interactions, as we discussed above, create images, and internal images serve to adapt a creature to its appointed place (§61). Signals and specificity go together; signals that are not specific cannot signify. Specificity is exclusivity; the ability to receive the 'right' signals incorporates the ability to exclude the non-signals, the 'wrong' signals. Specificity, therefore, defines your foothold in the world. What are the forces that create specificity? How far can they be trusted to define your world of interaction?

Let us consider some of the important details of signal recognition between a protein that acts as a receptor (that which receives) and a molecule it recognizes, which is termed a *ligand* (that which is bound, from the Latin *ligare*, to bind).

Recognition is a well-used word, and needs no explanation when we use it to refer to the abilities of people, or even to those of animals. We recognize friends in a crowd and our car in a parking lot; we recognize opportunity or danger. The dog recognizes its master and its home. But what does it mean to say that an unconscious entity, a physiological system or a molecule recognizes? In principle, recognition requires the fulfilment of two conditions: *discrimination* and *response*. Discrimination marks the specificity of recognition. One recognizes a specific face (or a molecular signal) when that face can be distinguished from other faces in the crowd. The detection of specificity is the detection of information (§7). We have defined the response to information as its meaning (§8). Recognition, as I wish to define it, requires both information and meaning.

One might argue that specific discrimination alone should suffice to define recognition; why do I insist on a response? Doesn't the recognition, in effect, precede the response? I have included the response in the act of recognition because there is no way to detect a recognition event unless there is a response event. Recognition leads to some change – a smile, a wagging tail. Indeed we observe recognition within our minds by observing our internal, mental response. To recognize a face (or a molecule) implies knowing what it (the face or molecule) means; some response to the signal constitutes the signal's meaning (§8, §66). A signal that means nothing, that elicits no change, is not a signal. Recognition is the association of an *exclusive signal* with some type of *response*. I use the term response in its broadest sense; a response may be negative, an inhibition or suppression of action and not only a positive activation. Recognition is thus akin to knowing, as we have defined knowing above (§44, §90), as a way of responding to the world. In summary, any thing – person, machine or molecule – that discriminates and responds can be said to recognize. Information and meaning are the ingredients of recognition.

§100 Biologic Recognition

Molecular discrimination is typified by the image of the interdigitating lock and key (§58). Because of their complementary shapes, the designated lock and key fit specifically; foreign keys that don't fit are excluded. A close fit is also meaningful because touching leads to binding, and binding makes possible interaction. The lock opens in response to the key because, at least for a moment in time, lock and key are bound together by the friction of their specific fit. Biological recognition too can occur when molecules fit and bind. Therefore, to appreciate how molecules may recognize one another, we have to consider both molecular fit and molecular response.

Molecular specificity is created by two types of forces: the forces internal to molecules that establish their shapes (and hence their capacity to fit one another) and the external forces between molecules that bind them together, the molecular friction so to say, that allows them to interact. We are interested in temporary binding because molecular recognition, like a handshake, is reversible. Reversibility allows the subjects to maintain their individuality despite their interaction.

101 Protein Receptors: Combining Sites and Reaction Sites

The receptors of the immune and other living systems usually are made of protein because a variety of different shapes can be constructed from proteins more readily than from other families of biological molecules. The shape of a molecule is called its *conformation* (from *form*), and proteins excel at allowing the creation of complex conformations. The diversity of possible protein conformations is so great that no ligand of importance need go unseen; some protein, such as an antibody, could be constructed in a way that would allow the protein to bind it. Diversity of shape is particularly useful at the site of the receptor molecule that binds the ligand, the *combining site* of the receptor. Diverse combining site structures ('locks') are needed to recognize a diversity of individual ligands ('keys'): a combining site for each ligand.

The second useful attribute of receptor proteins is their responsiveness; a protein can respond to a ligand by changing its conformation when it binds the ligand. Proteins are *relatively* stable, not absolutely stable. A protein may be able to assume more than one conformation with fidelity. The beauty of the matter is that the alternative conformations of a receptor can reflect whether or not the receptor has bound a particular ligand. The part of the receptor that responds functionally to the ligand can be called the *reaction site*. The reaction site can have one distinct shape in the absence of a ligand and another distinct shape when the combining site has bound (has sensed) a ligand. This shift in conformation of the reaction site constitutes the physical response of the receptor to ligand binding. Receptor proteins can thus be said to *recognize* specific ligands (§99) because they combine *specific conformation* (information) with *responsiveness* (meaning).

Receptor–ligand interactions can be viewed as *attractors* (§19). Consider the stable alternative shapes of a receptor protein as alternative basins of attraction. The binding of a ligand to the combining site of the receptor can push the reaction site of the receptor from one basin of attraction into another. Indeed, the attractor concept can supply us with a definition of a ligand; a ligand is a molecule that, through binding, can affect its receptor's conformational basin of attraction. Many sticky molecules may bind to a receptor protein, but only those that effect a response or a change in response are true ligands. See Figure 24.

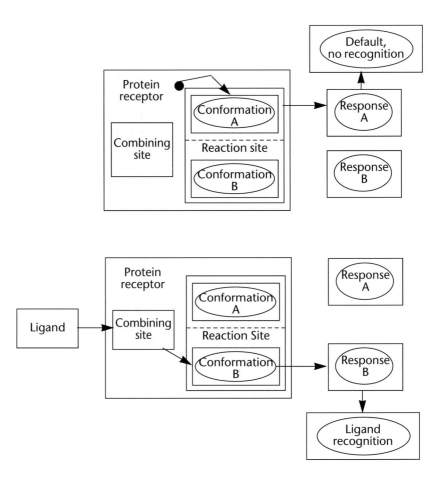

Figure 24: Ligand binding to the combining site determines the conformation of the reaction site and, hence, the response

The Figure illustrates a *protein receptor* composed of a *combining site* for *ligands* and a *reaction site* that mediates receptor function. The reaction site is shown in two of its possible stable conformation states (attractors), designated *conformation A* and *conformation B*. The upper frame of Figure 24 shows the receptor in the absence of a ligand. When the combining site of the receptor is not occupied by the ligand, the forces intrinsic to the receptor protein, designated by the internal arrow, maintain the reaction site in conformation A. In conformation A, the receptor's reaction site generates a particular response, *response A*. Since response A is not triggered by a ligand, we can call this response the *default*, or *no recognition* response. (In fact, the default response may include the absence of a particular response.) The lower frame of Figure 24 shows what happens when a specific ligand binds to the combining site of the receptor protein: now the reaction site is driven to assume a new shape, conformation B. In this shape, the reaction site of the receptor protein shifts its state of activity to response B. Response B effectively reports the act of ligand binding and triggers its consequences. Response B signifies ligand recognition. The ability of a protein to shift reaction conformations, A to B and B to A, breathing, as it were, in-and-out, is the protein's contribution to the variable interactions of life.

Note that a single protein may actually deploy a number of different reaction sites, each with its own function. And the activations of these functions may be responsive to diverse ligands, each interacting with its own combining site. Many enzymes, for example, can carry out different types of reactions depending on molecular signals and other factors in their environment: ligand signals interact with the enzyme's various combining sites and activate the needed reaction sites for the job at hand. The specificity of combining-site and reaction-site interactions allows us to map our worlds with fidelity (§58).

102 Mechanics of Protein Shape: Genetic and Epigenetic

From whence comes this complexity and responsiveness, the capacity to shift basins of attraction, to breathe? The answer, as you would guess, is complex. The factors that determine a protein's specific conformation, and hence its conformity to certain ligands and its responses, are *covalent* and *non-covalent* forces. Covalent bonds are formed between atoms when the atoms share electrons. A collection of atoms bound together by covalent bonds forms a molecule. Covalent bonds are 'irreversible' because atoms that share electrons do not usually diffuse apart, unless energy is applied to disrupt the connection.

Proteins are constructed out of chains of amino acid sub-units. There are twenty different natural amino acids that, by different combinatorial associations, can be strung together to make different proteins. The diversity of possible protein types emerges from this combinatorial strategy. Each of the amino acids in a protein chain is covalently bound by an *amide* bond to its neighboring amino acids. Like the pieces in a children's construction set, amino acids 'snap' together with a uniform amide bond element irrespective of how the amino acid sub-units might otherwise differ chemically. The order in which the amino acids are strung together, the amino acid sequence, is encoded in the DNA genetic code of the protein (§41). Thus the diversity of the genetic code creates the initial diversity of proteins.

The covalent backbone of a protein is only one of the factors that shape the protein. In contrast to the stereotypic regularity of the amide bond snap that forms the protein backbone, each of the twenty amino acids differs from its fellows in the size and chemical nature of the part of the amino acid called the *side chain*. Depending on the identity of the amino acid, its side chain can be smaller or larger, or more or less water soluble, or with or without a positive or a negative electric charge. These features of amino acid side chains make it possible for the amino acids in the sequence of the protein to interact non-covalently, without sharing electrons.

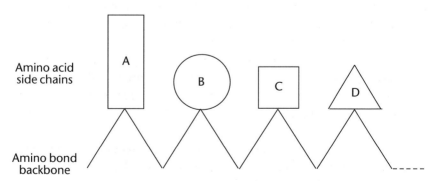

Figure 25: Amino acid chain

The amino acids in a protein (designated A, B, C, D . . .) are covalently bonded together to form a chain. The chemical characteristics of the side chains of the amino acids (designated by the different shapes) allow them to interact by non-covalent forces, leading the chain to fold into a particular shape.

There are four types of non-covalent forces that can mutually attract or repel the side chains of amino acids:

1. Ionic bonds – side chains of opposite electric charge attract, and those of like charge repel, like magnets.
2. Hydrogen bonds – hydrogen atoms, which act as if they are positively charged, can attract atoms such as oxygen or nitrogen that are electronegative.
3. Hydrophobic forces – just as oil will not dissolve in water, hydrophobic (water-hating) side chains will avoid water and seek the company of their similarly hydrophobic fellows.
4. Van der waals forces – at certain distances, atoms attract each other due to their fluctuating electrical charges. But when too close, atoms repel each other. These forces of attraction and repulsion establish the optimal distance between adjacent atoms.

The force of a non-covalent interaction works only at relatively close range and is weak, less than one-twentieth of the force of a strong covalent bond. Yet, weak non-covalent forces determine the conformation of a protein because they produce attractions and repulsions between the different side chains of the amino acids in the protein. The covalent backbone of the protein responds to these non-covalent interactions of the side chains by bending and by swiveling about its axis. Thus the protein will fold and twist this way and that in response to the non-covalent forces acting between the side chains. Indeed, non-covalent forces can also operate between adjacent strands of the protein's backbone. The most stable conformations of a protein will be those manifesting relatively low potential energy. Quite simply, conformations of lower potential energy are more energetically stable than are conformations of higher potential energy (§9).

Classically, the three-dimensional shape of the protein and, hence, its function were considered to be programmed by the one-dimensional information encoded in the DNA sequence of the protein's gene (see the view of Myer, §41): the gene encodes the protein's amino acid sequence, and the side chains of the particular amino acids, in turn, determine the protein's conformation. The stable conformations constitute the protein's natural attractors (§19–§22).

But here's the rub. The protein chain encoded by a single DNA sequence can have more than one stable conformation (and hence more than one function), and transitions between conformations are determined by ligands and other environmental factors (see Figure 24). By itself, the DNA sequence does not suffice to determine the shape and function of its protein; the shape and function of a protein emerge in response to interactions with the epigenetic environment (Greek *epi – outside of* the gene). In other words, three-dimensional protein conformation emerges (§12–§18) from a combination of genetic and epigenetic factors. The structure and the function of a protein arise from the one-dimensional information inherent in the amino acid sequence, which is encoded in the DNA gene, and from the physical energy of side chain and backbone interactions (§7, §9) supplied by the environment. Moreover, lymphocytes, as we shall see, may construct their own genes epigenetically (§111). The importance of epigenetics clearly challenges classical genetic determinism; see, for example, R. C. Strohman (1997).

The conformations assumed by proteins can be stable because the multitude of non-covalent forces between amino acid side chains, although individually weak, can combine to create a strong cohesive force. Consider the zipper: the friction that holds each claw to its neighbor is negligible – zipping or unzipping is easily accomplished by uniting or separating the claws one by one. But try and pull the jacket apart if the zipper gets stuck. Another example (macroscopic) of collective non-covalent force (microscopic) is the resistance of plastic wrappers to urgent fingers. Strength emerges from togetherness, weakness from disunion.

§103 Emergence of Protein Conformation

We have zoomed into the interior of the protein to uncover the genetic and epigenetic forces, covalent and non-covalent, that determine protein conformation. Thus the shape of the protein can be reduced to fundamental chemistry and physics. Protein folding illustrates the explanatory power of scientific reduction; we have in hand a physical mechanism adequate to fully explain protein structure. Nevertheless, protein folding also illustrates the weakness of classical reduction (just as it illustrates the weakness of classical genetics; §102).

The energy landscape of the interactions underlying protein folding is so complex that we cannot, even with the aid of a powerful computer, predict from a knowledge of the protein's amino acid sequence the exact shapes into which the protein will fold. When asked to compute the conformation of a protein based on its sequence, the computer flounders; there are simply too many computations that need to be done. The position of every amino acid side chain can affect the position of every other side chain; hence, any computed adjustment of a side chain forces the computer to make endless computations to determine the new position of every other side chain in the protein. That's why, ultimately, the exact conformation of a protein cannot be predicted from its sequence but has to be measured directly and laboriously using x-ray crystallography and other sophisticated physical methods, which hold, of course, only for the particular protein conformation that has been crystallized. It is a wonder how the protein, within seconds, folds itself into stable states of potential energy (§9). Obviously, proteins, unlike computers, do not have to compute their steps one by one as they fold. But then how do they do it? Again, the mystery of emergence (§12).

104 Chaperones, Stress and Maintenance Proteins

Although much remains to be learned about the folding process, we do know that a specialized class of proteins functions to help other proteins fold themselves properly. Proteins are synthesized and enter the cellular environment as unfolded linear chains. Protein synthesis is inherently hazardous, therefore, because the contents of the cell could become exposed to incomplete protein chains. Incomplete protein chains lack the amino acids they need for their mature conformations and so might assume improper shapes and, even worse, poison the cell by interacting illicitly with important intra-cellular structures. The problem of illicit interactions is particularly evident in periods of cellular *stress* (such as high temperature, dehydration, energy starvation) in which many of a cell's folded mature proteins may become unfolded, or *denatured*.

(If you wish to see denatured protein, you need only boil some milk and observe the appearance of a membrane on the surface as the milk cools; the film of scum, disgusting to most children, emerges from the denaturation of milk proteins that were soluble before they became unraveled, unfolded by the heat of boiling. Not all denaturations need be distasteful: fine cheese, for example.)

To control such potentially dangerous interactions, nature has evolved a special class of proteins that can bind to unfolded protein chains, isolate them from components of the cell and help them fold properly. These helper proteins are called *chaperones* because they prevent unacceptable liaisons. Chaperones can be viewed as a kind of pre-emptive ligand; 'Stick close to aunty, and you'll stay out of trouble', one might say. Some chaperones are also called *stress proteins* because of their critical function in protecting the cell during protein denaturation induced by stress. Indeed, stressed cells can be characterized by an increased production of stress proteins.

Stress proteins can be termed *maintenance proteins* because they function to maintain the organism in states of emergency (§98). Particular emergencies, being accidents, are unpredictable; as a class, however, emergencies are inevitable: we live in the dominions of chaos (§25) and entropy (§9). It is no wonder then that all cells in this world, from the most primitive of bacteria to the neurons of your brain, must be able to make stress proteins. No wonder, too, that stress protein genes have been among the most conserved genes throughout evolution. Protein denaturation threatens humans no less than bacteria, and stress protein genes, once they successfully evolved, have been irreplaceable.

You might guess that the immune system has developed a keen interest in stress proteins. To carry out its maintenance tasks, the immune system must be alert to cellular stress, and the appearance of stress proteins signals stress with great fidelity. Actually, stress proteins and the immune system, despite differences in their scales of operation, are partners in maintenance (§98). We shall discuss immunity to stress proteins and to other maintenance proteins below when we discuss *autoimmunity* and the immune system's internal image of the self (§149).

105 Specific Ligand Binding

Now that we have clarified the properties, genetic and epigenetic, of receptor proteins, let us return to recognition and discuss the molecular basis of binding site *specificity*. Molecular specificity hinges on the array of non-covalent bonds formed at the points of contact between the ligand and receptor. Ligand binding, therefore, is a natural extension of protein folding; they both result from similar non-covalent forces (§102). Indeed, a sufficiently flexible ligand, responding to different non-covalent forces in different binding sites, can mold itself to fit different receptors, a phenomenon called *induced fit*.

Note that receptor–ligand binding, like folding, is inherently reversible; the ligand can diffuse away from the grasp of the receptor. This reversibility makes occupancy of the receptor binding site sensitive to the concentration of the ligand. A certain concentration of ligand is required to keep the receptor binding site occupied in the face of spontaneous ligand dissociation.

The sensitivity of ligand binding to ligand concentration allows us to quantitate binding *affinity*, the energy of binding between the receptor and the ligand. If the non-covalent forces of attraction between receptor and ligand are great, then even at a low concentration of ligand, dissociation is unlikely and the binding site of the receptor will be occupied by ligand. Conversely, low affinity interactions lead to high rates of ligand dissociation, and so require a high concentration of ligand molecules to maintain occupancy.

Molecular specificity, therefore, is a matter of relative affinity. And affinity is never all-or-none; affinity is a matter of degree. This is an important point; let's consider what it means. Our macroscopic experience with receptors is the lock and the key, or the socket and the plug. Now keys and plugs seem either to fit their locks and sockets or they don't; you can't use your European electrical appliance in an American hotel (and vice versa) if you have forgotten to pack the adapter. The non-covalent bonding of ligands to receptors at the molecular scale is much more permissive.

§106 A Figure of Ligand Binding

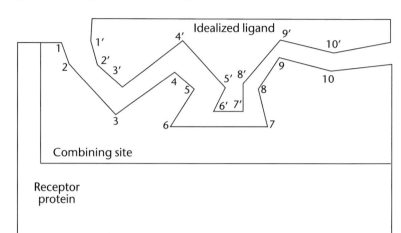

Figure 26: A schematic ligand-combining site interaction

Figure 26 is a two–dimensional, idealized cross-section of what is really a three-dimensional interaction between a ligand and the combining site of a receptor protein. The representation is hypothetical and the scale of the partners has no basis in fact. The drawing has been made only to show that a combining site has a number of possible contact points, numbered here arbitrarily 1–10. Each of the ten might establish non-covalent bonds with complementary contact points on an idealized ligand, numbered 1′–10′.

Let us imagine the possibilities of interaction between the receptor-combining site illustrated in Figure 26 and various ligands. Figure 27 charts a series of ligands that vary in their complementary contact points that might match the ten possible contact points of the schematic combining site of the receptor. The contact point interactions are known to be additive; and, to simplify matters, let us assume that their binding energies are equal.

Ligand A is ideal because it meets the binding sites of the receptor point for point. However, it is easy to imagine a less than ideal ligand, one

Ligand	1'	2'	3'	4'	5'	6'	7'	8'	9'	10'
A	+	+	+	+	+	+	+	+	+	+
B	−	+	+	+	+	+	+	+	+	+
C	+	−	+	+	+	+	+	+	+	+
D	+	+	−	+	+	+	+	+	+	+
E	+	−	−	+	−	+	−	−	+	+
F	−	−	+	−	−	+	−	−	+	−
•										
•										
•										
n	−	−	−	−	−	−	−	−	−	−

Figure 27: Complementary ligand contact points

that lacks a single contact point, and so makes one less non–covalent bond; Ligand B for example. Ligand B could also be imagined to bind, albeit with somewhat lower affinity than Ligand A. Likewise, hypothetical Ligand C would also bind. In fact, one could imagine a series of 1,024 (2^{10}) alternative ligands (A, B, C, D, E, F . . . n). One thousand and twenty-three could have at least one contact point and some measurable affinity for the binding site. Ligand n represents the set of non-ligands, those that cannot enter the binding site cleft or that can enter, but have no possible contact points. The other 1,023 ligands can be imagined to fit the binding site to some degree.

So wherein lies the specificity of recognition; how can a receptor exclude the various alternative ligands that might fit it? The point is that it can't; ligand–receptor binding is intrinsically leaky. The specificity of recognition, therefore, cannot rest on the initial binding event. If we want to see specific recognition, we shall have to look beyond. Actually, the leakiness of the initial binding event is a good reason, possibly the best reason, for insisting that the definition of recognition include a response (§99). As we shall see, specificity is not a given; the immune system must manufacture it.

DEGENERACY, PLEIOTROPIA, REDUNDANCY, RANDOMNESS

Specific recognition is essential for effective behavior, yet immune receptors are degenerate, pleiotropic, redundant, and random. These characteristics of the immune system negate any simple one-to-one relationship between cause and effect. How can the system generate specificity out of the basic non-specificity of its component parts?

§107 Degenerate by Nature

Figures 26 and 27 are greatly simplified; in reality, the problem of specificity is even more complicated. The range of affinities of ligand–receptor interactions can differ in their magnitudes by a million-fold and more, and yet interactions at the low end of the affinity scale can still transmit biologically effective signals. This means that a single receptor might actually be able to accommodate not merely 1,023 ligands, but possibly many more thousands of different ligands with some measurable affinity. The conclusion is inescapable; a combining site of any protein receptor intrinsically is able to bind many different ligands. There is no way that non-covalent forces acting between ligands and receptors could ever be exclusive to one pair of interactors. The word *fidelity*, like the word *pure*, is absolute only in concept. In biology, both words are always relative. Any receptor will bind more than one ligand, and, conversely, any ligand will be able to interact with more than one receptor.

The failure of receptors to be faithful to one ligand has been a nagging disappointment. Immunologists had wished to reduce specific recognition to a discrete ligand–receptor interaction (§141). Scientific reduction feels most confident when it comes upon a one-to-one relationship between cause and effect. The law of gravity, the paragon of scientific causes, allows no ambivalence. Immunologists, and biologists generally, would like to have been able to explain the specificities of their observations by simple ligand binding, but they have been thwarted by the complexity of nature. Regret may have led to the technical term used to describe the less-than-discriminate nature of ligand–receptor interactions: *degeneracy*. Degenerate receptors are sometimes called promiscuous; mind you, no value or moral judgments are intended by these scientific terms. But disappointment, I think, seeps through the terminology. (If you think that degeneracy and promiscuity are neutral terms, you may see what your dictionary has to say about them.) Perhaps we should all be disappointed; the specificity of our world rests on non-covalent bindings, and the world and its bindings are degenerate.

108 Degenerate Games

However disappointing at first glance to purists, the intrinsic degeneracy of receptors has its consolations. Degeneracy, for example, allows receptor–ligand interactions a degree of plasticity, and plasticity eases regulation. Consider the following: a uniquely specific ligand would have to be the ligand with the highest affinity, like Ligand A depicted in Figure 27. However, very specific, very high-affinity ligands often function poorly and may even be dangerous. As I said above, ligand–receptor interactions, like an effective handshake, should be reversible (§100). In fact, different receptor–ligand interactions are best served by different ranges of affinity, including low affinity. Enzymes, for example, need to become free of a present ligand in order to proceed to work on the next ligand. So the ligand with the highest affinity may not do the best job. Actually, a high-affinity ligand can kill you by sticking too tightly to its receptor. Some types of nerve gas, for example, work by having an affinity for a nervous system enzyme (acetyl cholinesterase) that is orders of magnitude greater than the natural acetylcholine ligand of the enzyme. The highly specific, high-affinity nerve gas ligand inactivates the enzyme by preventing it from interacting with its natural ligand acetylcholine. Even the most loving of couples needs room to breathe.

Nature, and not only the army, exploits high affinity for chemical warfare. Some snake venoms contain high-affinity ligands that kill by displacing lower-affinity natural ligands. Sea snails of a certain type are known to poison prey (and unwary divers too) by a blast of small proteins with high affinity for key molecules.

High-affinity disruption of natural ligand interactions can also do good. Some antibiotics work as high-affinity ligands to poison bacteria. In fact, pharmaceutical agents of all kinds work as artificial ligands to activate or to block receptors that have evolved naturally to bind other molecules (§6). Drug therapy, as we now use it, would not be possible if receptor–ligand interactions were more precise. Thus the very plasticity of receptor–ligand pairing can be a blessing.

Ligand plasticity has a functional terminology. An *agonist* (from the Greek word for *contest*, or *struggle*) is an alternative ligand that can activate a receptor in a way similar to that of the natural ligand. For example, the effects of the drug morphine are due to its abnormal activation of receptors specific for natural molecules (endorphins) produced by the body to regulate pain and pleasure; morphine is an agonist for endorphin receptors. (Agonists like morphine can save lives, when used as medicines, or ruin lives, when abused as narcotics.)

An *antagonist* (to struggle against) is a ligand that binds physically to a receptor; but rather than activating the receptor, the antagonist neutralizes the effects of an agonist. The antagonist can function as a blocking, or competing molecule; the antagonist occupies the binding site intended for the agonist and prevents activation of the receptor.

Altered ligands can act as agonists or as antagonists or, more intriguingly, as partial agonists. Imagine that Ligand B in Figure 27, which lacks a contact point (1′), does not induce the same conformational change in the receptor's reaction site as does Ligand A, but rather induces an alternative conformation of the reaction site. Now the conformation of the reaction site determines the receptor's function (see Figure 24, §101). Therefore, Ligand B (an alteration of Ligand A) could produce a different type of response than does Ligand A, although both ligands A and B bind to the same receptor-combining site.

These different functional definitions of ligands are not sharp, and there is much overlap between the classes. But the names are not really important. The important point is that the intrinsic degeneracy of the receptor-combining site provides rich opportunities for the regulation of recognition by alternative ligands. But that comes later (§12, §156). Before we discuss how the system works, let us continue to list the obstacles that seem to compromise immune specificity.

109 Pleiotropia

Immune agents are not only degenerate in what they sense, they are pleiotropic in what they do. *Pleiotropism* is a term denoting the capacity of a single agent, cell or molecule, to produce many diverse effects. *Tropia*, in Greek, means 'to turn on' and *pleio* means 'more than one'. The word 'pleiotropia' is free of the emotional undertones of *degeneracy*; but pleiotropia, nonetheless, obstructs one-to-one specificity.

The immune system is pleiotropic to an extreme. Cells are complex creatures, so it is not surprising that one cell can do many different, even contradictory things: a single T cell can kill one target and yet stimulate the growth of another. Immune pleiotropism, however, is expressed even at the scale of a single molecule. Pleiotropia does not refer to the alternative reaction states of a molecule (§101), but to a true diversity of function. Consider, for example, the cytokine called IFNγ (the designation for <u>interferon</u> gamma), which is produced by T cells and NK cells (§92). IFNγ, like the other classes of interferons (IFNα and IFNβ), was so named because the molecule was first discovered for its ability to *interfere* with the ability of viruses to infect cells. But IFNγ does not attract much attention now for viral interference; most interest in IFNγ arises from its ability to activate a variety of destructive immune effects. IFNγ, for example, is an angel when we want macrophages to destroy tuberculosis organisms; but it is a devil in destructive autoimmune diseases like type I diabetes or multiple sclerosis.

As we said above, cytokines work by activating genes (§93). Perhaps the different effects of IFNγ are more apparent than real; perhaps IFNγ activates only one gene which happens to express itself differently in different target cells. The basic question then is not what IFNγ seems to cause at a macroscopic scale, but how many genes does IFNγ activate at the molecular level. Jonathan Howard and his associates have actually done the tabulation, and it turns out that IFNγ activates more than 200 different genes (see U. Boehm, T. Klamp, M. Groot, and J. C. Howard, 1997). Each of the 200 genes are more or less pleiotropic themselves. So a single molecule like IFNγ is pleiotropic at the most basic causal level.

(Parenthetically, we might note that the emergence of pleiotropism is probably guaranteed by the laws of evolution [§31]. Complexity tends to accelerate during evolution because existing information and interactions provide further opportunities for the organization of new information and new interactions [§32, §72, §73]. Once a molecule like IFNγ becomes available, chances are that evolution will find a way to exploit it for new interactions – as long as the added complexity is stable. Good molecules, like good ideas, continually get reincarnated. Life is basically modular.)

Dozens of different cytokines have been discovered, and many, probably all, will be found to manifest pleiotropic effects. Indeed, many of the same cytokines initially were given different names according to the different guises of the cytokine that, by chance, attracted the attention of one or another research group. The cytokine known as TNFα, for example, was named *tumor necrosis factor* by a group interested in the killing of tumor cells (we now know that TNFα can trigger apoptosis in various types of cells). Another group studied what they called *cachexin*, a factor that caused body wasting (*cachexia-cacos*, *bad* and *hexis*, *state*, in Greek), resulting from an inhibition of appetite and an increased metabolism of body fat. The two, apparently different factors turned out to be pleiotropic effects of the same TNFα molecule.

To complicate matters further, not only may a given cytokine activate seemingly unrelated effects, but different cytokines may produce very similar effects. We now know, for example, that many of the effects of TNFα, but not all, overlap with those of IFNγ. The functional overlap between different immune agents raises another barrier to one-to-one specificity, that of redundancy.

§110 Redundancy: Simple and Degenerate

We discussed redundancy above when we defined its importance as a safety factor in the process of self organization (§73). Redundant means superfluous or extra; a redundant effect is one produced by more than one agent. We can distinguish between two types of redundancy: simple and degenerate. Simple redundancy denotes the existence of multiple copies of the same agent. Simple redundancy is important for the process of self-organization because simple redundancy preserves information in the face of change. But simple redundancy is trivial when we consider specificity. It matters little to the specificity of a receptor just how many receptor molecules there are. (It does matter if the extra copies of the receptor have to compete for a limited amount of ligand, but that's another story.)

We can apply the term *degenerate redundancy* to describe the situation in which several different agents perform the same action. Although the agents are not identical, they are redundant to the degree to which they may replace one another functionally. If, for example, we observe an immune effect that could have been produced either by IFNγ or by TNFα, we cannot be certain which of the cytokines was responsible, unless we study the effect. Degenerate redundancy compromises the one-to-one specificity we would have hoped to see between a particular effect and a particular cause.

Degenerate redundancy is further complicated by the fact that different agents can produce some of the same effects, but also different effects. For example, TNFα and IFNγ only partially overlap in their immune effects; both cytokines amplify destructive inflammation, but each cytokine produces effects that the other does not. T cells, NK cells and macrophages are redundant inducers of apoptosis, but they are completely dissimilar in other of their actions. We may thus define degenerate redundancy as the partially overlapping pleiotropisms of diverse agents.

111 Random Combining Sites: Antigen Recognition

We said above that T cells and B cells were princes (§92). Perhaps we should have called them wizards. T cells and B cells can manage a unique and wondrous trick that no other cells in the body can match; they can create receptor genes somatically.

Except for two families of proteins (and for rare mutations that can occur throughout your DNA), all the amino acid sequences of all the proteins in your body, as far as we know, have been produced according to DNA sequences inherited from the human germ-line and transmitted to you through the union of your parents. The two exceptions are the antigen receptors of your T cells and your B cells (including the antibodies secreted by the B cells). You do not inherit the DNA genes that encode your antigen receptors; you manufacture your own receptor genes epigenetically from genetic raw materials.

Recall that receptor proteins include two functional sites: a combining site, which binds and senses the ligand, and a reaction site, which acts in response to the ligand binding (§101). In an antigen receptor, the genetic information for these two sites differs in origin. We receive from our parents germ-line DNA sequences that encode the receptor reaction sites, but it is up to our lymphocytes to construct somatically the DNA that encodes the ligand–combining sites. The combining site part of the receptor DNA is called the *variable* region because it varies; each T cell and B cell can construct a different combining site. In contrast, the part of the receptor DNA that encodes the reaction site is called the *constant* region because the DNA encoding that region is expressed as encoded in the germ–line, without somatic modification.

How do lymphocytes construct variable region receptor genes? It is done by rearranging a number of relatively small segments of germ-line DNA. There are three families of different gene segments called V, for variable, D, for diversity, and J, for joining region genes. And each of the families contains from five to 70 or more different DNA segments. To create a functional antigen-combining site, each progenitor T cell or B cell randomly recombines one DNA segment from each of these three DNA families. The sequences of DNA that intervene between the rearranged segments are spliced out and discarded. This combinatorial rearrangement creates a large number of potential V-D-J combinations.

(Combinatorial rearrangements are also effectively exploited by human culture: consider the numbers of different words you can generate by combinatorial rearrangement of the 26 letters of the alphabet. The diversity of proteins too emerges from amino acid combinations; §102.)

The construction of antigen receptors goes a step beyond simple combinatorials. The splicing of the V-D-J segments is engineered to create mutations, deletions and additions to the DNA sequence at the segment junctions. (Some receptor chains recombine V and J segments without D segments, but that does not much change the picture.) Hence, each of your T and B cells can express an individualized combining site that may never have been expressed until now in any human antigen receptor. In other words, the 'master DNA plan' itself is edited epigenetically and permanently in each lymphocyte to generate uniquely variable antigen-combining sites. All this work, as we said, is somatic. Your children and children's children will have to make their own antigen receptors from scratch. (If you happen to be female, you may help your children somatically with your own antibodies, but more on that later; §128.)

Although you cannot transfer your private antigen receptor gene repertoire to your children, your T cells and B cells can transfer their antigen receptor genes to their daughter cells (for some reason, cells give rise to *daughter* cells, and not to *son* cells). The antigen receptors that are generated during lymphocyte differentiation are passed down to the offspring of the progenitor lymphocyte. Thus, each progenitor and its offspring constitute a unique clone that is marked by its unique antigen receptor. Your repertoire of T cells and B cells is a large collection of individualized clones.

To construct a functional receptor, it remains for the lymphocyte clone to splice the joined V-D-J genes (or the V-J genes) to one of the several alternative constant-region gene segments that form the reaction site of the receptor (or antibody). In this way, your antigen receptors are contrived to express somatically variable combining sites (to sense the variable world of antigens) along with constant germ-line reaction sites (to report to your body the meaning of the encounter) As we shall see, the variable V-D-J combining site of an antibody can be stitched to different constant reaction sites to generate different responses to the same antigen (§118). Your clones have options.

(You might recall our earlier discussions of *decisions* [§54] as *associations* between *particulars and classes* [§55]. We may consider the somatically generated combining site as a *particular* [the combining site senses particular antigens], and the germ-line response site of the receptor as encoding the *class* of response appended to particular antigens [§75]. Thus, the double-jointed construction of the antigen receptor is a molecular representation of the double-jointed immune experience of both the species [germ-line] and the individual [somatic; §91]. But, patience, we are getting ahead of our story.)

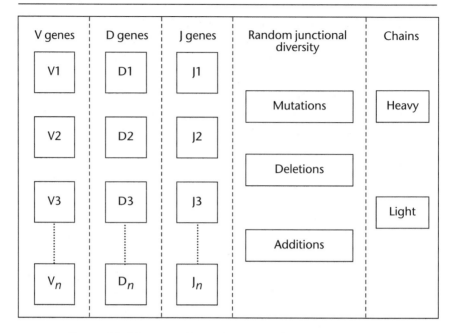

Figure 28: The genetic engineering of antigen receptors

Figure 28 shows the combinatorial strategy of the T-cell and B-cell antigen receptors.

The gene segments, which are spliced together to form the combining site of the antigen receptor, belong to several gene families designated V, D (in some situations), and J. Each gene family may contain from five to 70 or more different segments. Combinatorial splicing of the various V, D and J segments produces a large potential diversity of combining sites. For example, each of 50-odd V genes could combine with any of 20-odd D genes, and the resulting VD combinations (let's say about 1,000) could combine with, say, five different J gene segments to produce 5,000 different V-D-J sequences. Now each receptor is formed by random combinations of two different chains (light and heavy chain families), each of which is the product of independent recombinations. Therefore, the thousands of possible sequences in one chain must be multiplied by the thousands of possible sequences in the other chain. This gives rise to some millions of different potential V-D-J antigen-combining sites that could be attached to a constant region segment that encodes a reaction site. Moreover, the random mutations, deletions and additions of nucleic acids, which materialize at the splicing junctions, create an added potential for further variable region diversity. It is estimated that about 10^{11} (a hundred billion) different B-cell receptors (and their antibodies) could be produced, and perhaps 10^{15} different T-cell receptors (a million times a billion). Your body contains about 10^{12} lymphocytes, each of which expresses only one (or possibly two) receptors; so only a small fraction of your potential receptor diversity gets expressed at any one time.

Nevertheless, even the small fraction that is produced is a very large number of diverse receptors. Your body contains only about 100,000 other types of proteins, each inherited from the germ-line of the species by way of your parents. Thus the epigenetic machinery of an individual's lymphocytes is able to produce about a millionfold more diversity than does the germ-line of the species. Indeed, your immune system is as rich in the potential diversity of its antigen receptors as is your brain in the number of its nerve connections. By combinatorial machinations, your lymphocytes have the potential to see an untold number of different molecular conformations; any molecule could become an antigen.

Now that we have defined the antigen receptors, we can finally define an *antigen*: an antigen is any molecule that can be recognized by an antigen receptor. The part of an antigen that actually fits into the receptor is called the *epitope* of the antigen. Defining an antigen by an antigen receptor may sound like a circular definition, but it's true to life; the immune system operationally defines its own somatic world of ligands.

The somatic creation of astronomical numbers of different antigen receptors using random combinations of relatively small numbers of V, D and J segments is a unique triumph of the immune system; the discovery of the process is a triumph of immunology (see *The Generation of Diversity. Clonal Selection Theory and the Rise of Molecular Immunology*, by S. C. Podolsky and A. I. Tauber, 1997). But how practical is such largesse? As we discussed, each of the millions of different antigen receptors are degenerate; each can sense many different ligand antigens (§107). How can the immune system make effective use of such a large potential? How can it hope to extract specific signals from the noise that must be generated by so many receptors?

We can look at the problem this way: information is the resolution of uncertainty (§7). But the primary repertoire of antigen receptors is the ultimate expression of random uncertainty. Hence, the primary receptor repertoire, of itself, bears no information. Specific recognition, which requires both information and meaning (§99), will have to be resolved, as we shall soon see, through the progressive self-organization (§73) of the system and its receptor repertoire.

§112 Specificity Enigmas

Let us summarize the problem of immune specificity. All biologic interactions involve transfers of information (§6), and all involve, in some form or other, the binding of ligands to receptors (§100). Yet, specificity of discrimination cannot be achieved at the level of a single receptor–ligand interaction. We have devoted many words to elaborate this point, going all the way down to the atomic scale, because degeneracy is a cardinal fact of life; life is the expression of ligand–receptor interactions that are intrinsically lax. In other words, functional biologic specificity cannot be explained by an underlying chemical or physical specificity in any *one-to-one* ligand–receptor binding relationship. The immune specificity problem is further aggravated by three additional snags: pleiotropia, redundancy, and randomness. Immune agents can produce a multitude of overlapping effects, and the ability of the lymphocytes to sense antigens is random and effectively unlimited.

In short, biologic specificity cannot be reduced to the chemistry and physics of ligand binding, but must be contrived by biologic mechanisms. Signals are always circumstantial. Hence, specificity cannot be a given but must be a creation. Despite the fundamental laxity of receptor–ligand binding, immune behavior is functionally specific; witness the ability of the immune system to distinguish between the parasite it attacks and the host body it protects. The immune system does, as it must, discriminate right from wrong. The emergence of immune specificity out of chemical degeneracy is the process we must now explain. In the coming sections, we shall build specific immune cognition in three stages, discussing the *geometry* of cognition, or cognitive *space*, the *dynamics* of cognition, or cognitive *time*, and the *images* of cognition, or cognitive *meaning*.

GEOMETRY OF IMMUNE COGNITION AND CO-RESPONDENCE

Cognition arises, at the ground level, from two spatial properties of immune agents: the different agents detect different features of the antigenic world, and the agents mutually co-respond to form regulatory networks. Macrophages sense the context in which the antigen appears (the state of the tissues and the presence of infection); T cells analyze a part of the amino acid sequence of the protein antigen bound to a molecule of the major histocompatibility complex; and B cells recognize the conformational shapes of antigens. These immune agents reinforce and modify their diverse views of the world (co-respond) by way of germ-line and somatic regulatory networks involving cytokines, antigen receptors, and by-stander collectives of antigens. The immune system responds to its own response.

§113 Feature Detection: Focused Perception

The immune system can, in theory, express a random repertoire of millions of degenerate antigen receptors. Such a repertoire would be capable of recognizing any imaginable antigen. Yet, to see everything at once is to court chaos. Specificity requires focus – the active extraction of a signal from a background of potential noise. Attending to certain features while ignoring others is an efficient way, probably the only way, to funnel a potential flood of input into organized channels of guided experience. To be useful, the potentially unlimited capacity of the antigen–receptor repertoire must be constrained. Wise parents counsel fidgety children to ignore distractions and focus their attention on priorities. As we discussed above, the brain uses in-built feature detectors (§62) and attention preferences (§63) to fashion specific experience out of raw sensation (§76). The immune system uses a similar strategy of feature detection. Let us examine macrophages, T cells and B cells, and note the focus of their recognition preferences.

§114 Macrophages Sense Context

Macrophages express germ-line receptors, and so recognize germ-line molecules. Unlike lymphocytes, macrophages do not somatically create receptors, and so cannot recognize antigens. Nevertheless, the germ-line features macrophages do see greatly influence the somatic choice of antigens recognized by T cells and B cells. Macrophages note and report to the lymphocytes the *context* in which the lymphocytes recognize their antigens. We can divide the contextual perceptions of macrophages into three classes: the state of body tissues, the presence and effects of infectious agents, and the state of activation of nearby immune agents.

Tissues: macrophages resident in specific body tissues express properties suited to the maintenance of the particular tissue. The cytokines produced by activated macrophages differ in different parts of the body; the brain, the eye, the skin, the gut, as we pointed out, are serviced by different types of macrophages suited to the characteristic challenges in maintenance and protection confronting the tissue (§94). Resident tissue-specific macrophages differentiate from the pool of circulating monocytes by responding to local tissue signals. In addition to their ability to discriminate between various healthy tissues, macrophages are able to sense tissue damage because they bear receptors for ligand molecules exposed when cells are damaged and receptors for the cytokines produced by damaged cells. Signs of cell and tissue damage attract circulating macrophages to join the resident macrophages at the site, and the macrophages are activated to destroy aberrant cells and remove the debris. By secreting a series of cytokines, the macrophages help initiate the processes of scar formation and healing. The macrophage cytokines also activate the cytokine receptors of lymphocytes.

Infection: macrophages can sense infection because they bear receptors for molecules unique to infectious agents, such as components of bacterial cell walls or carbohydrate molecules characteristic of bacteria. Macrophages recognize virus particles too. Molecules of infectious agents activate macrophages to attack, engulf and kill the invaders, and to process invader molecules into antigens for T cells (§115). The responding macrophages express cell-interaction molecules and cytokines that report the infectious context to T and B cells. The lymphocytes are thus activated by the macrophages to respond vigorously to antigens associated with infections.

Immune activity: macrophages sense the cytokines and other signal molecules produced by lymphocytes. Macrophages bear receptors (called *Fc receptors*) for the reaction sites of antibodies (the *Fc domain*), which makes it possible for macrophages to recognize antibodies bound to antigens. Bound antibodies flag macrophages to engulf and destroy the targets of the antibodies.

Imagine, for example, a smart bacterium or virus that attempts to evade detection by hiding its ligands (distinctive cell-wall or carbohydrate molecules) for which macrophages have germ-line receptors. But have no fear, a molecule of the parasite for which the macrophage has no receptor can still be converted into a targeting signal for macrophages. An antibody to that parasite molecule does the trick by tacking its reaction site (Fc) domain onto the parasite (by way of the antibody's combining site, of course). The macrophage simply recognizes the bound antibody's Fc domain as a surrogate marker for the parasite, and kills it. Thus, the somatic antibody and the germ-line macrophage work together to bridge a gap in germ-line recognition, and so outwit an evasive parasite; germ-line macrophages see with somatic–antibody eyes.

We can now define *immune context*: context is the array of ancillary signals that influence the response to a specific subject. Your response to a man with a gun differs greatly depending on the context, whether the act has taken place on the stage, or in the subway on the way home from the theater.

Of course, the distinction between context and subject is relative to one's point of view. The macrophage, unlike the lymphocyte, 'views' the antigen (marked by the bound antibody), not as its subject, but only as a contextual signal for attacking the infectious agent, the macrophage's true subject. The lymphocyte, in contrast, 'views' the macrophage as part of a context for seeing the antigen, the lymphocyte's true subject. The drawn gun, the subject of your attention, is only the gunman's contextual prop for the subject of making a living through acting, either on the stage or in the subway.

§115 T Cells, the Major Histocompatibility Complex and Processed Antigen

Lymphocytes recognize antigens; what features of antigens are preferred by T cells? Your potential repertoire of T-cell antigen-combining sites is enormous (§111), but is sagely bounded by significant restrictions. The combining site of a T-cell receptor interacts, not with an antigen as such but with a ligand composed of two independent sub-units: one sub-unit is a piece of an antigen molecule and the other sub-unit is a part of a self molecule of the major histocompatibility complex (MHC). Figure 29 clarifies what I mean by a compound ligand.

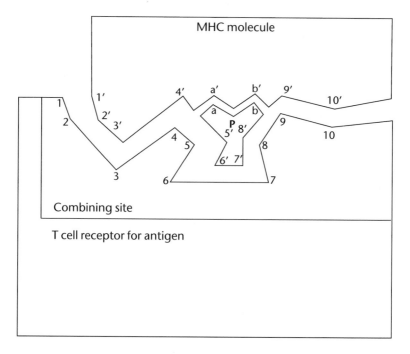

Figure 29: The T-cell receptor recognizes a compound ligand

This schematic picture is a modified copy of Figure 26 (§106). The combining site of the T-cell receptor is shown with the same hypothetical ten contact points represented in the earlier picture. The ligand in Figure 29, however, has been modified. The ligand still has ten possible complementary contact points (1'–10'), but now the contact points are distributed among two separate ligand entities: the MHC molecule itself furnishes points $1'$–$4'$ and $9'$ and $10'$, and a peptide (**P**) connected to the MHC molecule furnishes contact points $5'$–$8'$. All the combining-site contact points are there, but now they come from more than one ligand molecule.

Note another difference between Figure 29 and Figure 26: the compound ligand itself shows internal contact points. Peptide **P** has points, a and b, that are in contact with complementary points, a' and b', in the MHC part of the compound ligand. Thus a compound ligand results from a tandem interaction: a full ten-point interaction of the T-cell receptor with the **P**-MHC compound ligand cannot be satisfied unless there has been a preceding contact-point interaction of **P** with the MHC molecule. (To simplify the picture, the contact between **P** and the MHC is shown with two non-covalent points, a and b. In reality, there are usually between two and four major contact points that anchor **P** to the MHC molecule).

Note that the **P**-MHC interaction is the *context* in which T cells recognize antigens. Like all contextual points of reference, the tandem interaction of **P**, the MHC, and the T-cell receptor has important consequences for the T-cell repertoire's window on the antigenic world. The enormously diverse T-cell receptor repertoire is designed to detect subtle structural variations on a theme of defined features. Let us analyze the features of the **P**-MHC interaction context that must precede the entry of the T cell into action.

You probably know of people, perhaps even yourself, who suffer from allergies to 'unnatural' substances such as medications, chemicals or even metals (the chromium or nickel in one's wedding ring, for example). T cells are involved in some of these allergies, which indicates that T cells have some way of recognizing 'unnatural' substances. For the most part, however, T cells recognize fragments of proteins, peptides like **P** in Figure 29 (in fact, it is likely that allergenic drugs and metals activate T cells by first modifying peptides, but that is not important for now). The proteins that can give rise to suitable **P** peptide fragments may originate from within a body cell, from a true self-protein molecule or from a foreign molecule encoded by an infectious agent resident in the cell. Such internal **P** peptides tend to be presented to T cells in association with a family of MHC molecules called MHC-I.

The important point is that almost all the cells in your body produce MHC-I molecules, so all the cells in your body can be scrutinized by your T cells. In fact, your T cells can examine any protein made in any of your body cells, provided that at least one peptide fragment of the protein (**P** in Figure 29) can be bound by an MHC-I molecule.

Now why did I say '*provided that at least one peptide . . .* '? The reason is that certain conditions must be fulfilled before an antigen peptide can form a complex ligand with an MHC-I molecule: the protein antigen has to be susceptible to the protein-degrading machinery of the cell; the resulting peptide fragments of the antigen have to bear signals that allow their transport to the MHC-I molecules in the cell; the peptide molecule has to be the proper size to fit the binding site of the MHC-I molecule; and, finally, the peptide has to have contact points (like *a* and *b*, for example in Figure 29) that allow it to be anchored non-covalently to complementary contact points (*a′* and *b′*) in the peptide-binding domain of the MHC-I molecule. These contextual requirements markedly limit the types of peptides that can form **P**-MHC I ligands on the cell surface. Only when all these conditions have been satisfied can the T cells have a look at the **P**-MHC-I complex. If macrophages can be said to see infectious agents with the help of somatic antibody eyes (§114), then T cells can be said to see antigens through germ-line (MHC) spectacles (provided by macrophages and other antigen-presenting cells).

In addition to the **P**-MHC-I system used for presenting internally produced peptides, there exists a family of MHC molecules, called MHC-II, that specializes in presenting processed peptides fashioned from proteins taken up from outside the cell, or made by the cell and expressed on the cell surface. In contrast to MHC-I molecules, which are expressed essentially on all of our cells, only special antigen-presenting cells (APC) produce MHC-II molecules; these include macrophages and other phagocytes that engulf and digest extra-cellular materials, B cells, activated T cells, and some tissue cells that are seriously stressed. Thus, MHC-II molecules are reserved for special immune communications.

Like the **P** fragments presented to T cells by MHC-I molecules, the **P** fragments presented by MHC-II molecules must fulfil certain molecular requirements in order to be bound to the MHC-II pocket. The details of antigen processing and presentation to T cells are beyond the scope of this book. The important points are that MHC-I and MHC-II molecules differ in their **P** peptides, and that different types of T cells specialize in docking to **P-MHC-I** or **P-MHC-II** compound ligands (CD-8 and CD-4 T cells respectively). Thus, different classes of peptide, by binding to different MHC molecules, declare their origin (from inside the cell or from the cell surface), manifest different chemical features, and activate different types of T cells.

A macrophage will present antigens differently through the MHC-I and MHC-II pathways, depending on the integration by the macrophage of various signals (tissue damage, infection, immune status). In response to these contextual signals, the macrophage can adjust its antigen-processing machinery, its expression of **P-MHC** molecules, and the quantity and amount of time the **P-MHC** is available on the cell surface. The macrophage can also present the **P-MHC** along with various types of cytokines and accessory molecules that activate or inactivate T cells or B cells (§119). Macrophages in this way report the antigen context to responding T cells.

Note a very interesting, perhaps surprising feature of the T-cell world: most of the contact between the T-cell receptor and its **P**-MHC ligand is with the MHC part of the complex. Rather than recognizing pure antigens, the T cell actually responds to one's own MHC molecules as they are modified by a few protruding side chains (contact points) of a bound peptide molecule. For the T cell, which is the real subject and which is the context: the **P** or the MHC? Indeed, T cells can be activated by MHC molecules that don't even present a specific peptide fragment (**P**) to the T-cells' antigen receptors; a super-antigen is the term for any molecule that activates T cells by circumventing the antigen and linking the T-cell receptor to an MHC molecule directly. Look at Figure 29 again and imagine that the **P** fragment is replaced by a molecule that links the T-cell receptor to the MHC molecule outside of the combining site; such a linker is a super-antigen. Super-antigens activate large populations of T cells, not distinct T-cell clones with particular antigen receptors.

Ponder this: the marvelous molecular machine that fabricates the enormous diversity of T-cell antigen receptors has evolved to note small structural modifications of the self-MHC molecule. Why so much for so little? A great investment usually signifies a great need. We shall consider this point below (§132, §156).

§116 MHC Polymorphism and Sexual Preference

A single MHC molecule, either class I or II, can present many different peptides to T cells, provided that the various peptides share the common binding motif needed to anchor them to the MHC molecule (like *a'* and *b'* in Figure 29, for example). A peptide that lacks the *a'–b'* motif will not be able to be presented by that MHC molecule to T cells. Therefore, despite a repertoire of millions of T-cell receptors, the individual will be immunologically blind to any protein that lacks the required peptide motif for the given MHC molecule. Such a restricted view of the world could leave the T cell blind to important antigens. How can the T-cell repertoire gain a wider world-view? The answer is simple; provide the T cells with several different MHC molecules, each with different anchor points for different peptides. This, indeed, is what the immune system does. Each one of us inherits from our parents three different MHC-I genes and three different MHC-II genes.

Note that we are born with two separate copies of each gene: one from father and one from mother. Thus, if father and mother bear the same MHC-I and MHC-II genes, we gain nothing, at least in this regard, from having two parents. However, if mother and father have each provided us with different alleles of the MHC genes, then we inherit twelve different MHC binding sites, instead of only six, and considerably increase the diversity of the processed peptides visible to our T cells. (Alternative versions of a single gene are called alleles; from the Greek *allos* – 'other'.) Therefore, to double your T-cell world-view, the species need only make sure that your father and mother carry different MHC alleles.

But how can the species be sure that couples will be likely to pass on different MHC alleles (so that you will have twelve different MHC molecules rather than only six)? There appear to be at least two mechanisms for increasing your chances of being born to parents with different sets of MHC genes. First, the MHC genes are extremely polymorphic (*many forms*); there are a very large number of different MHC alleles dispersed among members of the species so, if mating is random, prospective parents will be likely to carry different alleles by chance.

Mating, however, is rarely random; most people express strong mating preferences. What's to keep individuals from choosing to mate with other individuals who bear the same MHC alleles, thereby MHC short-changing their offspring? Amazingly, it turns out that having a different MHC is sexually attractive, at least to mice.

One would not suppose that mice think very much about beauty, economic prospects, or intelligence before they mate; they seem just to do it, at least whenever the female is in estrus (prepared hormonally for mating). Nevertheless, mice prefer, when given a choice, to mate with partners who express MHC molecules that differ from their own. In addition to encoding the presentation of processed peptides, MHC alleles appear to encode particular body odors. These odors, detectable by other mice, render the bearer sexually attractive (or repulsive). The selective mating of individuals that differ in their MHC alleles will reinforce a couple's chances of passing on to their offspring two sets of different MHC alleles rather than only one.

We don't know yet if people, like mice, are attracted to partners with diverse MHC genes. But who knows? How often have we wondered what he/she could possibly see in her/him? When all other explanations fail, consider the irresistible tug of a novel MHC allele. (Should we tell the perfume industry about a new opportunity, or just leave MHC attraction to nature?)

MHC polymorphism is serious business when it comes to the rejection of transplanted organs, as we shall see (§156). MHC differences also seem to be important in the immunological relationship between a mother and her fetus (§128).

117 NK cells Guarantee MHC-I Revelations

We said earlier that NK cells can sense and kill abnormal body cells (§92). Now that we know about the MHC and antigen presentation to T cells, we can be more explicit about the abnormality sought by the NK cell. NK cells kill body cells that do *not* express a particular MHC-I molecule.

The logic goes like this: MHC-I molecules present to T cells peptides processed from internal proteins. The **P-MHC-I** system, as we said, makes it possible for the T-cell repertoire to survey a sample of the molecules made by the body's cells (§115). Now a cell that is not expressing an MHC-I molecule is a cell that is able to hide its proteins from patrolling T cells. In fact, certain viruses inhibit an infected cell's production of MHC-I molecules and, consequently, these infected cells cannot present virus peptides to immune T cells. Some tumor cells also stop making MHC-I molecules. Without MHC-I spectacles, T cells are functionally blind to a cell's antigenic abnormalities. A body cell that has stopped making MHC-I molecules could constitute a danger to the rest of the body by hiding its aberrations from patrolling T cells.

The NK cell comes to the body's rescue. The species has learned that it is better to kill cells lacking MHC-I molecules than it is to spare them and be sorry later. And so the dutiful NK cell presses the apoptosis button of body cells that should be showing their MHC-I molecules, but don't. A healthy cell, like a virtuous soul, should have no secrets to hide.

NK cells, like macrophages, also have receptors for the Fc domains of antibodies, and can kill antibody-marked targets.

In summary, NK cells don't see the antigens of body cells, but they make sure that the body cells are honestly exposing their inside antigens, through the **P-MHC-I** machine, for T-cell inspection. Death is inflicted, again, for life (§97).

§118 B cells and Antibodies Recognize Molecular Conformation

In contrast to T cells, which recognize denatured peptide fragments of proteins, B cells and the antibodies they secrete recognize the conformations of proteins and other antigen molecules. The secreted antibodies serve as specific molecular agents of the mother B-cell clone.

The immune effects of an antibody depend on the combining site of the antibody, which sees the antigen conformation, and on the reaction site (Fc site) of the antibody, which triggers various immune effects. There are about a half-dozen different antibody reaction sites (called antibody isotypes), and each reaction site isotype dictates the type of immune effect the antibody can cause. Some antibody isotypes activate phagocytosis better than others; some activate complement enzymes to kill targets; some are secreted more readily at the interfaces of the body (the guts, the lungs, the genital tract); some are passed on by the mother to her fetus (§128), and so forth. Each antibody isotype has its specialty. How does the B cell decide what kind of reaction site it should append to the combining site of its particular antibody? T cells tell the B cell which reaction site to deploy.

Soluble antigens taken up by B cells (via their antigen receptor) can be processed and their peptide fragments complexed with MHC-II molecules and presented to T cells. The T cells, in turn, can respond to the P-MHC ligand and to particular context signals by secreting different cytokines. The T-cell cytokines activate the B cell to attach various constant region antibody-reaction sites (antibody isotypes) to its antibody combining site. In the absence of instructions from T cells, a B cell will produce antibodies with a default type of reaction site (IgM isotype). Thus, a B cell, on its own, can recognize an antigen conformation; but to exercise a choice of response type, the B cell needs a cytokine signal from a T cell. The T cell, as we shall discuss below (§119), decides by integrating contextual signals what kinds of cytokines it had best send to the B cell.

119 Co-Respondence

We are now prepared to discuss a key property of immune geometry alluded to earlier: co-respondence (§91). Macrophages, T cells and B cells, working in parallel, note different features of any target entity in need of immune maintenance or protection, be it an antigen, an infection, or a damaged cell or tissue. A macrophage sees the contextual features, a B cell sees a conformation of some antigen and a T cell sees a sample of an antigen's amino acid sequence (in an MHC context). Most importantly, each of the three agents not only makes its own observations, it tells the others about its response. The mutual exchange of information is not a mere courtesy; each of the three agents updates or changes its own response in the light of the responses of its fellows. Immune agents, in effect, respond not only to antigens and other target objects, they also respond to the response of fellow agents to different features of the target object. The overt immune response seen macroscopically is thus an integration of a series of covert, microscopic responses to the effects of the target on the system, as well as to the target itself. The immune system, in short, responds simultaneously to different aspects of its target entities and to its own responses to these target features. We may call this mutual exchange of signals *co-respondence*. The involved population of macrophages continues to perform its germ-line program according to the tissue context, the T-cell population continues to analyze processed peptides, and the B-cell population sees antigen conformation and secretes antibodies; but each agent does its act with more or less vigor and with different effects as it co-responds with its fellow agents. Co-respondence is committee action, a type of communion, between different agents declaring different perceptions.

Figure 30 summarizes the features detected by macrophages, T cells and B cells, and their mutual influences through co-response signals; cytokines, cell interaction molecules, processed peptides, antibodies. The response of each type of agent modifies the responses of the other agents, each to its unique world of perception.

Co-respondence is a seminal concept. We defined recognition as both seeing and responding (§99). Recognition, from the viewpoint of co-respondence, is subject to committee deliberations and decisions. The specific response to an immune situation is thus a matter for consultation among immune agents. Recognition, at the macroscopic level, emerges as a property of co-responding populations of semi-independent agents.

Co-respondence also tells us something about the basis of immune specificity. Biologic receptors, as we discussed, are degenerate intrinsically; any receptor can bind more than one ligand (§107). No single receptor is trustworthy. Co-respondence allows specificity to emerge in a courtroom of immune deliberations. The macrophage, the T cell and the B cell are witnesses to different aspects of an immune situation. Co-respondence is a type of mutual cross-examination in which the perceptions of the different immune agents (each faulty to some

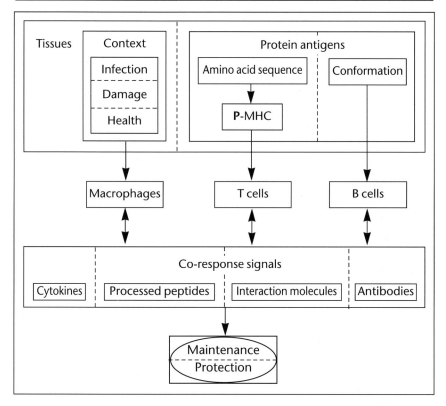

Figure 30: Co-respondence

degree of degeneracy) lead to joint immune behavior, the macroscopic immune response. Note that there is no external judge that rules on the deliberations; the co-responsive interaction itself is what creates the court's decision. The interaction of the semi-independent agents is the court of specificity. Truth is thus the compound image of an interaction. Immune truth, like other types of truth, is conditional, not absolute. Immune specificity is a collective of interacting features; change some of the perceived features and you change the specific response. Specificity is not given at the start; specificity is the outcome of the response (§106).

Co-respondence can even help us grasp some of the logic of immune anatomy. Lymph nodes can be viewed as the courts where immune agents can gather to present their findings for communal co-respondence *in camera*, secluded from distractions in the tissue arena of action. Once their mutual deliberations lead to a joint immune decision, the agents can exit the lymph node to return to the blood for delivery of their verdict to the tissues. Lymph nodes are not only district offices (§95), they can function as local, *ad hoc* brains.

An important feature of co-respondence is that the tissues of the body also take part in the process (§140). The tissues produce many of the same signal molecules (cytokines, adhesion molecules) produced by the immune agents. Let us consider the means by which the immune agents (and the tissues) discuss their perceptions and co-respond.

§120 Cytokine Networks: T1 and T2

At the level of the germ-line, it is possible to divide various cytokines and the cells that produce them into groups marked by common redundant effects. T-cell cytokines have been divided into two polar groupings: T1 and T2. The T cells and the cytokines of the T1 set tend to activate, directly or indirectly, destructive immune effects. The T1 cytokine group includes IFNγ, TNFα, IL-1, and IL-12 (we have discussed IFNγ and TNFα above [§110]; the IL acronym stands for *InterLeukin, leukocyte interaction* molecule. Numerical IL designations are now used to name cytokines in place of the historical acronyms rendered obsolete by the discovery of cytokine pleiotropism).

A second group of T cells is the T2 set, which produce cytokines like IL-4, IL-5, IL-10, and TGFβ. T2 cells and their characteristic cytokines, in contrast to the T1 set, tend to activate immune effects that are less destructive, and often promote healing.

T1 and T2 cytokines are produced by macrophages, B cells and tissue cells in various states, and not only by T cells. Cytokines are key signal elements in co-respondence. The member cytokines of each set not only manifest redundant pleiotropic effects, they actually stimulate each other's production. For example, IL-12 stimulates IFNγ and vice versa, and IL-10 and IL-4 mutually reinforce one another's production by T2 cells.

Most important, reinforcement within the sets is accompanied by opposition between the sets; T1 and T2 cytokines tend to suppress both the activities and the cytokine production of the other T-cell set.

However, as if to thwart our quest for one-to-one causality, it turns out that the pleiotropic effects of individual cytokines may also be contradictory; in some circumstances, members of the destructive and healing camps may switch roles. TNFα is a notorious T1-type destroyer of target cells, but it also stimulates healing by triggering the growth of blood vessels. IL-4 belongs to the T2 healer group, but it can contribute significantly to destructive autoimmune diseases like systemic lupus erythematosus. Moreover, many cells do not fit into the polar T1-T2 dichotomy, and can produce variable amounts and mixtures of typical T1 and T2 cytokines. The interactions between the cytokine sets are complex. Administering the same cytokine to an experimental animal at different stages of an immune response can lead to opposite effects. IFNγ, for example, is a classical T1 cytokine. Yet IFNγ treatment can abort inflammation, or enhance it, when given at different times.

The complexity of cytokine interactions is compounded by the participation of cytokines and cytokine receptors in regulatory circuits. For example, the receptor for TNFα acts as a cell's sensor for TNFα when the receptor is anchored to the membrane of the cell. However, the TNFα receptor can also be released from the cell membrane and circulate in the body fluids in a free, unanchored form. In its free form, the TNFα receptor can also bind its TNFα ligand, but in this situation, no TNFα signal will be transduced into a cell. On the contrary, the free receptor can now act as a competitor and, by binding TNFα, prevent it from stimulating the receptors anchored to cells. Thus the state of a receptor can regulate the function of its ligand. In fact, the TNFα ligand itself can be found in at least two different states: free, or anchored to cell surfaces. Anchored TNFα can function as an adhesion molecule. In other words, a ligand can sometimes act as a receptor, and a receptor can sometimes act as a ligand, depending on whether the molecules are anchored or free.

121 Feedback

Feedback is a factor in co-respondence. Feedback refers to the effect of a product on the product's production. Feedback can be either positive and enhance production, or negative and suppress production.

Eating, for example, is produced by appetite (among other factors). Eating an appetizer is aimed at positive feedback; a successful appetizer sharpens your interest in the coming courses. Eating the meal itself generates negative feedback; satiety annuls appetite and ends the meal. The positive feedback of our first-course appetizer ultimately leads to the negative feedback of the full meal on our appetite. So too can the interactions of T1 and T2 cytokines, and their receptors, provide both positive and negative feedback to the interactions of immune agents. Because of co-respondence, positive signals for one set of agents can generate negative signals for other sets of agents, and vice versa, leading to a chain reaction of signals. The generation of signals out of functional molecules enhances fidelity. Recall our discussion of the principle of the *receivers that transmit* (§64).

§122 Idiotypic Networks

We saw how the human eye both receives visual information from the world and transmits to other eyes information about its attentions (and intentions; §64). So too can the combining sites (the eyes) of antigen receptors function as ligands (transmitters of information) to the combining sites of other antigen receptors (other eyes).

Since the combining site of each T- or B-cell receptor is generated by DNA rearrangements that are combinatorial and random (§111), any combining site could act as a unique antigen capable of being recognized by another receptor. Indeed, the intrinsic degeneracy of receptors (§107) makes it possible for combining sites to sense other combining sites that might mimic the structures of potential antigen ligands (§108). Lymphocytes, in short, can recognize other lymphocytes to create internal networks.

The receptor-combining site, viewed as a ligand, is called an *idiotope* (*idio* in Greek means *one's own*, or *unique*, *topos* means a *site* or *marker;* an idiotype is a unique marker). Thus, receptor-combining site interactions are called *idiotypic*, and so give rise to idiotypic networks (idiotypic networks are sometimes called *anti-idiotypic*, but it makes no difference).

Look again at Figure 26, and now see the interacting elements as two interacting receptors, rather than as a single receptor and its ligand. T cells and B cells can see the eyes of other T cells or B cells and co-respond accordingly.

123 Antigen Collectives and By-Stander Effects

Two or more clones of immune system cells may be held in close proximity at a particular site because each cell may recognize a ligand present at the site. Different antigen epitopes may be packaged together as components of a single molecule, bacterium, virus or cell. Therefore, different immune cells will be brought together whenever they interact with ligands that are linked together physically. Lymph nodes, which trap antigens and other ligand signals from their immune districts, are convenient sites for collecting from the blood and lymph various immune cells and molecules that might recognize individual signals in the trapped collective (§95). The co-responses of the collected cells can affect the whole collective. Immune effects related to the physical proximity of immune agents are called *by-stander* effects.

By-stander effects may seem crude, but they can strongly influence collective behavior. Consider the infectious character of a peaceful neighborhood, or the madness of a mob. Indeed, committees of immune agents need to be in proximity to interact; lymphocytes lack the long-distance wires of nerve cells (§95). The only way that the different features seen by macrophages, T cells and B cells can be associated with a single entity is by their localization to a single place of cell interaction. The court, as we said (§119), needs a courtroom. Cytokine and idiotypic networks are also dependent on spatial contiguity. Immune geometry is necessary for immune co-respondence. Relations in space create immune interactions, specificity and decisions.

124 Immune Networks

Immune co-respondence thus depends on immune networks. A network is a system in which the components are connected to each other such that a change in the state of one component can have some effect on the states or connections of the other components. One can easily imagine extensions of molecular and cellular networks in the number and intricacy of their interactions. If we consider all the immune agents that can, by co-respondence, reinforce, oppose and feed back on each other, both positively and negatively, and we multiply their points of connection through the degeneracy of signaling (§107) and the pleiotropism of response (§109), then we can begin to appreciate the complexity of the networks that constitute the immune system. These networks are materialized through the anatomical networks of lymph nodes and other tissues connected by blood and lymph channels and punctuated by selective cell adhesion (§94). The immune system, in its basic reality, is a collection of more-or-less connected networks distributed over the entire body. The complexity alone inspires awe. Complex networking is the raw material from which cognition emerges. Networks establish the topography, the space, in which interactions can occur in time.

DYNAMICS OF IMMUNE COGNITION

The operation of the immune system in time is an ongoing process of self-organization of the immune repertoire. Immune dynamics involves four rounds of selection. The first round of selection takes place among T-cell clones maturing in the thymus; T cells with a degenerate capacity to recognize self-P-MHC ligands are selected to populate the body. The second round of selection takes place in the peripheral lymphoid organs as T cells and B cells acquire immunological experience with their antigenic worlds. T cells are activated by experience with altered ligands. The third round of selection occurs as B cells are selected for high affinity antigen receptors in the course of an immune response. The fourth selection involves the adjustment of the response repertoire of cytokines and other germ-line effector molecules to the context of immune needs. This selection of germ-line effects proceeds simultaneously with the three rounds of selection of the somatic antigen–receptor repertoire. Mother helps prime the selection process.

§125 Primary T-cell Selection

Above, we stressed two facts about the antigen receptors of T cells: the receptors are generated randomly and the receptors, in practice, recognize a compound ligand consisting of a processed peptide bound non-covalently to a molecule of the major histocompatibility complex (**P-MHC**; §115). Is there a contradiction here? How can a repertoire be randomly generated, yet restricted in what it sees? Furthermore, the restricted view must be functional; we want the T-cell receptor repertoire to contribute to the maintenance of the body and to its protection against foreign invaders. The repertoire, hence, should have a clear view both of the self and of the threats to the self. The question, then, is how the random repertoire gets itself structured so it can do its two jobs. Structure, quite simply, evolves through selection; or better, through cycles of selection.

The first formative selection takes place, for T cells, in the thymus. The thymus is the central lymphoid organ in which immature T cells sojourn and are selected to become functional adults (§94). B cells undergo a process of formative selection in the bone marrow, but much less is known about how this occurs.

Shortly after each T cell generates its unique antigen–receptor combining site, the newborn T cell is obliged to undergo a test of fitness. The test is a life-or-death trial of receptor binding. If the baby T cell is not activated by binding to a self-**P-MHC** compound ligand, the T cell is programmed to activate its death mechanism (apoptosis). Recognize, or die.

Note two points about this selection. First, the processed peptide, **P**, must originate from the self because the thymus is normally sequestered from contact with the outside world; all the proteins available there for T-cell interactions come from the self. Secondly, the affinity of binding has to be just right – neither too weak nor too strong. Binding that is too strong will kill the T cell just as surely as will binding that is too weak. Recognize just so, or die.

Remember our discussion of affinity, the greater the affinity between a receptor and its ligand, the greater the specificity; and, conversely, the weaker the affinity, the greater the degeneracy (§106). In the first, thymic round of selection, high affinity is lethal (§107). In other words, most newborn clones of T lymphocytes will be selected for life only if they are moderately degenerate for binding to the **P-MHC** ligands they meet in the thymus. Thus, it can be said that the primary selection of the cell repertoire is for T cells that are moderately degenerate recognizers of the proteins in one's own body. The system, in a word, selects for degenerately autoimmune T cells.

Is this not a strange way to organize an effective repertoire? Self-maintenance would seem to require receptors well suited to certain self-molecules, while protection against foreign invaders would seem to require receptors well suited to foreign molecules. What could be gained in starting with a repertoire degenerately specific for the self? Probably the best of both the worlds of maintenance and of protection. The logic for degenerate self-selection goes like this.

First, some selection must take place to weed out T cells that might bear T-cell receptors incapable of recognizing the MHC contact points in a compound **P-MHC** ligand (§115). Receptors that are generated by random recombination have no way of proving they can bind MHC contact points, other than by doing it.

Secondly, the evolutionary connection between all manifestations of life on earth guarantees that all forms of life share molecular similarities. The self is not really very different chemically from its potential predators. The foreign is functionally a chemical alteration of the self. Thus, an immune system that has been selected to see the body with sufficient degeneracy is an immune system that should be able to see at least some epitope (some molecular fragment; §111) of whatever parasite may come to invade the body. Self-experience is the only experience accessible to the immature immune system, and, fortunately, protection can arise from self-experience.

Thirdly, maintenance of the body requires some degree of self-recognition. So self-selection is an effective introduction to self-maintenance. By a mechanism that is not yet clear, some T- (and B-) cell antigen receptors are apparently selected to recognize certain self-molecules, including maintenance molecules like p53 (§98) and stress proteins (§104), with high affinity. We shall discuss these special autoimmune T and B cells later when we discuss autoimmune disease (§149).

In summary, the primary experience of immature T cells with self-**P-MHC** in the thymus fashions the initially random repertoire into a structured population of clones manifesting a degenerate, but defined degree of specificity for the self. And a transition from randomness into structure, as we know, is information and the beginning of meaning (§7, §8).

126 The Second and Third Selections: Altered Antigenic Experience

The first, primary selection of the immature T-cell (and B-cell) repertoire, which takes place in the central lymphoid organs (§94), can be seen as preparation for the future confrontation with the world. The second and third selections, which take place in the peripheral lymphoid organs, are the actual responses to the world. The first selection is guidance, home tutoring (§76); the second and third selections are the acquisition of street wisdom (§75). The antigens associated with bashed bones, scalded skin, renegade cells, invading bacteria, parasitic viruses, worms, strawberries, and intimacy (and pregnancy) are the selecting facts of life in the world.

Note that the selective experience acquired in the periphery differs in an important way from the primary selection of the repertoire. The T-cell clones selected in the thymus for *moderate* affinity recognition of self-P-MHC are activated vigorously by *high* affinity P-MHC ligands in the periphery. Rather than inducing apoptosis, as they did in the thymus, high-affinity ligands now induce the T cells to proliferate and to produce effector molecules, such as cytokines. The shift of affinity upwards means that a T-cell clone is likely to respond in the second selection to ligands that differ to some degree from the ligand that saved the clone from death in the first selection. If the first selection led to T cells with degenerate affinity for self, then the second selection leads to T cells with higher affinity for *altered* self (§108). Now what is a ligand that is an alteration of a self-ligand? A foreign, but related, ligand. Thus the second selection, at least that for T cells, appears to focus on deviations from the self. The foreign deviations, nevertheless, must be self-like. There are also selections for unaltered self-recognition; but we shall discuss those later (§152).

The third selection appears to be limited to the B cells. In the course of responding to antigen conformation, and depending on co-response signals from T cells, the responding B cells are triggered to mutate the

genetic elements encoding the combining sites of their antigen receptors (§101). Clones of B cells that happen to have generated combining
sites with high affinity for the particular antigen are selected to expand
their numbers and produce large amounts of antibodies of particular
isotypes (§118). In other words, responding B cells, with T-cell help,
can increase their specificity and decrease their degeneracy in the
course of responding to antigen conformation. Thus the B-cell repertoire, as it becomes street-wise, creates sharper pictures of its world.
The process is called *affinity maturation*. T cells do not undergo
affinity maturation when they respond; apparently the system needs its
T cells to stay degenerate. Why should the T cells persist in seeing a
fuzzy world, and why should the myopic T cells be needed to control
the antigenic visual acuity of the B cells? We'll have to consider that
later (§131).

§127   The Fourth Selection: The Immune Effectors

The fourth selection is one we know very little about, although it's a
critical process because it creates immune meaning. The first three
selections can be said to establish the perception repertoire of the
immune system; by selecting particular T-cell and B-cell clones, the
three selections determine the antigens (and perhaps other signal
molecules) that the system can see. The fourth selection establishes the
biologic effects that result from the perceived signals; here the system
decides not what it sees, but what it does. This selection determines
the actual immune effects of the response.

Antigen perception, as we noted, is a somatic capability based on randomly generated antigen–receptor combining sites. Maintenance and
protection, the outputs of the immune response, in contrast, are
performed by molecules encoded in the germ-line (§111). The five
interventions produced by responding immune agents (gene activation,
cell growth, cell death, cell movement, and supply and support) are
germ-line capabilities (§96). The fourth selection, therefore, involves
the interface between the somatic antigen–receptor repertoire and the
germ-line response repertoire that effects protection and maintenance.
The fourth selection determines which types of potential response are
to be connected to which perceived signals.

Different situations require different effector agents: trauma, stroke, infections with various bacteria and viruses, tumors, and so forth each require different modes of response. Moreover, a fit response to a particular situation changes with time; there is a time to kill the invader or tumor cell, and there is a time to heal the wound. The nature of an effector response must evolve over time as the response itself modifies the situation. The effector response to perceived signals must continuously be updated and accommodated to the circumstances. This fourth selection thus involves immune decision-making, the heart of cognition. The decision-making process, as we shall see, will bring us to consider the chemical language of the immune system (§137).

128 Mother's Help

One may argue about the relative contributions of mother and father to the early emotional development of their offspring. But there is no argument about the predominant influence of mother on early immune development. Human mothers transmit a sample of their antibodies to their children. The transmission takes place through two routes: the blood and the gut. Late in fetal development, the placenta transports maternal antibodies to the fetus. These antibodies not only passively supply the newborn baby with ready-made antibodies against environmental pathogens common to mother and child, the maternal antibodies actively prime the immune system. As we discussed, macrophages readily ingest substances that are flagged by bound antibodies (§114), and the macrophages, in turn, produce **P-MHC** ligands for selecting T cells (§115). Mother's immune experience thus guides the baby's choice of targets for early immune responses. Daddy may get up at night to give the bottle, but mother's milk naturally means more to baby.

Antibodies secreted into mother's milk enter baby's gut. These gut antibodies neutralize pathogenic viruses and bacteria that populate the baby after birth, and also influence the active immunity the baby develops against the pathogens. Ungulates (cows, sheep) transmit maternal antibodies through the gut only, and newborns that miss suckling the colostrum (antibody-rich milk) often die of massive gut infection. Humans, fortunately, get most of their maternal antibodies through the placenta.

A final word about mother, father and immune maintenance. It seems that fertilization is promoted by an inflammatory reaction in mother's genital tract stimulated by father's sperm. Maternal inflammation is also important in the implantation of the fertilized egg into the wall of the uterus. Mother's immune system helps make you as well as keep you.

SELF-ORGANIZATION OF IMMUNE COGNITION

Having considered the agents of immunity and their geometry and dynamics, we shall now discuss immune cognition, defined formally: the immune system embodies self-organization, internal images, and decision-making. We begin with immune self-organization.

§129 Immune Self-organization

Self-organization, as we said, is the progressive creation of new information (§71). Starting with random noise and redundancy, a self-organizing system generates just-so arrangements (§73). That's the essence of self-organization, and that's the essence of the immune way. The T-cell and B-cell repertoires are the paragons of somatic self-organization. The primary sets of antigen receptors are produced by way of random genetic mutations and recombinations (§111). Subsequent interactions with antigens of the self and the environment (the three selections; §126) generate a structured arrangement composed of selected T-cell and B-cell clones. This somatic structuring of the repertoire includes idiotypic networks (§122).

Self-organization of the innate germ-line arm of the immune system is no less important. The degenerate redundancy of cytokines (§110) and their pleiotropism (§109) make possible a great many different cytokine arrangements in response to any given immune stimulation. From this point of view, the realization of a particular cytokine profile in an actual response can be seen as the creation of specific information, the resolution of uncertainty regarding the pattern of the response (§7). The selection of a particular response from the potential response repertoire, the fourth selection (§127), constitutes true self-organization.

Perhaps the most telling argument for germ-line self-organization comes from gene 'knock-out' experiments. One way of studying the function of a gene is to destroy it (knock it out), and then see what happens to the mouse missing the gene. Logically, whatever befalls the gene-deprived mouse can be blamed on the missing gene. Curious immunologists have hastened to knock-out key cytokine genes such as TNFα and IFNγ (§120) hoping to discover yet more functions for these agents – IFNγ, as you will remember, is known to activate more than 200 different genes (§109). The surprising conclusion of many of the gene knock-out experiments is how little a gene knock-out may affect the system. The immune system, for example, can remain macroscopically functional despite the loss of IFNγ; the system still makes antibodies, rejects foreign grafts, even produces autoimmune diseases. There are noticeable defects, but the system still manages to work.

But, you might ask, if the immune system manages to function to any degree without a gene for IFNγ, how can one claim that IFNγ is an important cytokine? Your question is apt. It turns out that the immune system, indeed, can be severely crippled by the loss of IFNγ (neutralized by an anti-IFNγ antibody, for example) provided that the system has first organized itself in the presence of the cytokine. IFNγ is irreplaceable only to an immune system that has developed using the cytokine. IFNγ, however, does appear to replaceable, at least in part, by an immune system that has developed without it. In other words, the immune system normally learns to depend on IFNγ during its formative period. But the system can also organize an effective immune response using alternative mixtures of molecules to replace an absent IFNγ.

The contingent importance of a germ-line molecule such as IFNγ is *prima facie* evidence of germ-line self-organization. The germ-line effectors, and not only the somatic receptors, are subject to self-organization. Clearly, the immune system self-organizes at all levels; it can use different molecular means at the microscopic scale to build a very similar immune capability on the macroscopic scale.

IMAGES AND PATTERNS OF IMMUNE COGNITION

The cognitive immune system helps equip the individual for survival by making images of the individual's internal and external environments that are concrete, abstract and distributed. The concept of pattern is central. Specificity emerges from subtle patterns composed of degenerate and overlapping elements.

§130 Immune Images

The second attribute of cognitive systems is their ability to construct internal maps. A creature survives thanks to internal images that map critical features of the environment in which the creature has to live, as we discussed above (§61). Maps are adaptive; maps prevent blunders (§30). A map tells you in advance where you are going. A unique virtue of cognitive systems is their ability to learn as they go by making internal images out of somatic experience. One does not have to argue about the existence of brain images; the reader is invited to conjure them up for inspection. Our immune images may not be open to sensation, but they are open to logic. Immune images can be *concrete, abstract,* or *distributed.*

Concrete images. Just as locks are concrete negative images of the keys they bind (§58), antigen receptors are concrete negative images of the antigens they bind – compare Figures 26 and 29 to Figure 6. Anti-idiotypic receptors (receptors to other receptors) constitute positive concrete images of the antigen ligand seen by the idiotypic receptor (§122). These images are concrete in the sense that they are based on physical points of contact, like the lock and key in Figure 6.

Abstract images. Images, however, are not always constrained by static, physical space; images can be abstract. Any interaction between two or more entities can be seen by the mind's eye to create abstract images of the interacting partners (§58). Mutuality of information, as we discussed, allows interacting partners to fit one another. The interaction fit creates an abstract informational space that maps the interacting partners. Thus an immune reaction, the mix of cells and cytokines involved in the response, maps, in effect, the antigens and other signals that elicited the response. The cytokine response to a virus, for example, is a functional image of the virus, even though no cytokine binds or recognizes the virus; no virus, no response. A high white count can serve as a functional image of an infection, even though none of the white cells expresses the complementary physical 'shape' of the infection (§92). The doctor looks at your white count and announces that he 'sees' an infection. The doctor sees the white cells concretely; the immune image of the infection 'seen' by the doctor is abstract. A concrete image, such as a photograph, lasts as long as the physical representation persists; the image disappears when the photograph fades. An abstract image lasts as long as the interacting entities interact, or can potentially interact. A doctor's image of your infection lasts as long as it can be 'seen' by a doctor. Abstract images are not made of matter; they are created by processes.

Let me clarify an important point about process images. Although I have talked about abstract images 'seen by the mind's eye' or 'seen by the doctor', such images exist even if no brain or eye ever perceives them. Interactional images may be beheld, but their existence is not merely in the mind of the beholder. Interactional images are not somebody's interpretation; such images independently exist as long as the interaction exists, or could exist.

Distributed images. Images need not be localized to a particular molecule or cell in an isolated site, but may exist as patterns of various molecules or cells distributed throughout the body. Such patterns may be called *distributed* images. For example, your entire T-cell repertoire, the pattern of different T-cell clones resident in your immune system at a given moment, is a distributed image of all the T-cell selections that have taken place in your body until that moment. Recall our discussion of how the present pattern of a river maps the history of the river's past evolution (§28). A pattern of T-cell reactivities records much of an individual's immune history. Patterns, because of their complexity, can be the most subtle of distributed images. Patterns, as we shall next discuss, actually draw sustenance from degeneracy, partial redundancy and pleiotropism.

§131 E Pluribus Unum: Specific Patterns

The question of specificity has occupied us on and off from the moment we began thinking about signals and receptors (§99). How can the immune system maintain and protect us in the face of the degeneracy (§107), redundancy (§110) and pleiotropism (§109) of its agents? How can the system work efficiently in the apparent absence of one-to-one causality (§112)? The time has come to try and resolve the question. I would like to propose that the seeming defects in one-to-one specificity are not the problem; on the contrary, they are the solution. Degeneracy and pleiotropism turn out to be raw materials essential for the construction of effective immune patterns. The argument goes like this.

A diversity of images of an object can contain more reliable information about the object than can any single image of that object. For example, the larger the number of witnesses who identify a particular suspect in a police line-up, the more certain we are of the identity of the culprit. The more letters of recommendation we receive telling us about the candidate, the greater the reliability of our decision. In other words, a pattern of different but overlapping reports (degenerate redundancy) bears more trustworthy information and is more credible than is a single account.

(Let us avoid the formal complication of an image that is an exact replica of its object in every detail; let us agree that a precise replica of a thing is the thing itself, and not a descriptive image. We may consider an image as an imprint of the object registered somewhere else; an image may be viewed as a kind of response to the object. But that is beyond our present scope.)

Patterns are important; but how do we know a pattern when we see it? Is a pattern any arrangement that pleases the mind's eye? I suggest we pause and define a *pattern*. The word 'pattern' is apparently derived from the word *patron*, and carries the sense of a role model. A pattern is a particular arrangement of elements that constitute a model to be used or emulated. Patterns are patterns if they can do something, if they can cause something to occur with some regularity. Patterns are interactional images (§130). Therefore, a pattern is a pattern even if no person likes the arrangement, or even if no person sees it. Patterns, like other abstract images, don't need observers. Hence, we may define a pattern as an arrangement that expresses a reproducible and meaningful relationship between relatively independent components.

Information, as we saw, can be defined as an arrangement (§7); meaning, in contrast, is information that does something (§8). A pattern, therefore, is akin to information with meaning. One could think about the components of a pattern as partners in a basin of attraction; the components of a pattern are related by their interactions leading to a particular pattern outcome (§22).

I use the term 'pattern' here to draw attention to arrangements composed of a connected diversity of images. Being many-to-one arrangements, patterns can convey more specific information about a complex situation than can a one-to-one arrangement. Matters biological are complex, so biology thrives on patterns.

Note, in fact, that life is composed of patterns within patterns at multiple scales (§15). Consider that an amino acid or a nucleic acid molecule is a reproducible arrangement, a pattern of atoms (atoms themselves are patterns, but don't ask me of what); patterns of amino acids constitute sequences of proteins and patterns of nucleic acids constitute genes; patterns of non-covalent forces help create protein shape; patterns of protein shape create recognition; and patterns of recognition form immune systems and their behavior, which is what we are now discussing. Biologic patterns are processes, and continuously require energy to keep alive. Patterns are interactive arrangements. They dissipate, they succumb to entropy, unless they are worked on. (At the end we'll think about the patterns that form us.)

Let us now consider pleiotropism from the pattern aspect. A multiplicity of different effector agents that can each respond to a given situation in pleiotropic ways offers more response options than does any lone (one-to-one) effector agent. A range of different combinations of degenerate and pleiotropic agents can create rich and varied image patterns capable of mapping diverse situations. If there were only one T1 cytokine and only one T2 cytokine, we would have specific one-to-one cytokine causality, but only of two polar types of cytokine response. However, if we have four, five or more partially overlapping T1 or T2 cytokines, as we actually do, then we can have a multitude of different T1 and T2 cytokine patterns to serve us, and not only two. Indeed, pattern diversity is amplified by combinatorial mixtures of the various T1 and T2 cytokines. The availability of many diverse image patterns makes it possible to assign a different pattern for each of a large number of subtly different situations. Each pattern of degenerate and pleiotropic agents (molecules and clones) can be highly specific for each variant situation, despite the degeneracy and pleiotropism of the individual agents. On the contrary, it is the very degeneracy and pleiotropism of the agents that make possible the diversity of specific patterns. Specific patterns emerge from populations of degenerates.

§132　Pattern Colors, B Cells and T Cells

We can illustrate the specificity of degenerate populations introspectively. Look about your world; how many colors can you distinguish? Certainly hundreds, likely thousands. (We happily buy costly computer screens advertised to show us thousands of colors.) Well, how many different color receptors (cones) in your retina provide your brain with thousands of different colors? Only three. I repeat; only three. One red, one green and one blue. Your brain can 'see' thousands of different colors because different combinations of signals from the three cones can form thousands of different specific signal patterns in your brain. The many patterns (colors) emerge from just three cones because the cones are somewhat degenerate (each cone responds to some degree to many different wavelengths of light) and partially overlap (the same wavelengths of light may trigger more than one type of cone, to some degree).

An artist paints a multitude of color patterns on a canvas by using his or her palette to mix various proportions of a relatively few primary colors. Your eye, using just three types of degenerate and overlapping cones, collects some of the photons bouncing off the canvas, and transmits the signals to your brain. The brain, as it were, uses its own palette to mix the signals and, thereby, recreates thousands of patterns (colors) for your enjoyment of the picture. Imagine what you would be missing if each cone were absolutely specific for only one photon wavelength and there were no overlap. You would see only the number of colors for which you have corresponding cones. You would see no intermediate colors. No degeneracy and no overlap means fewer possible patterns. And few patterns means poor discrimination. And poor discrimination means limited specificity. (Obviously, total degeneracy and total overlap won't make patterns either. Degeneracy and overlap, and redundancy and pleiotropism have to be graded to allow an optimal number of patterns; but that's beyond the scope of our present discussion.)

The concept of specificity-in-pattern can also help us understand how a mouse can manage an immune response with its IFNγ gene knocked out (§129). Most color-blind people lack one of the three types of cones; all the red cones are knocked out, for example. Nevertheless, people manage quite well with only two cones. Color-blind people, and the people around them too, often fail to detect the color-blindness, unless the person's color vision is tested for some reason. Apparently, even two types of cones can transmit a reasonably functional number of color patterns to the brain. So too can the remaining T1 cytokine types make enough patterns to compensate, at least in part, for the missing IFNγ in the knock-out mouse. Cytokine specificity, like color specificity, emerges from patterns, not from one-to-one arrangements.

The specificity of patterns is demonstrably evident in antigen recognition by antibodies. The early hopes of immunologists to the contrary, populations of identical antibodies derived from a single clone (monoclonal antibodies) are usually less specific than are mixed populations of different antibodies (polyclonal antibodies). A monoclonal population of antibodies is intrinsically poor in ligand specificity because the population is uniformly degenerate (§106). Polyclonal antibodies tend to be much more specific because each of the different antibodies features a different pattern of degeneracy. Each antibody in the polyclonal population sees its own degenerate world. The one ligand the different antibodies do recognize in common is the particular immunizing antigen that activated the heterologous population of antibody-producing B cells. The individual degeneracies of each antibody-combining site are thus diluted into insignificance within the polyclonal population of different antibodies. The degenerate bindings are functionally hidden. Antibody specificity thus emerges (§12) from a population pattern. There is greater specificity at the scale of the mixed population than there is at the scale of any single component antibody. (Democratic pluralism is usually more effective in defining social specificities than is monolithic authority.)

Above, we noted that T-cell recognition is not only degenerate, it is physically quite minimal; most of the contact of the T-cell is with the MHC component of the P-MHC compound ligand (§115). T cells thus recognize variations on self-MHC themes. Wherein lies their specificity for antigens? T-cell specificity, like specific color perception, emerges from the varying patterns of T-cell populations. T-cell antigen recognition begins with the degenerate structures of individual T-cell receptors, but this recognition gains specificity through the organization of patterns of T cells at the scale of the population.

But why, you might ask, does the immune system need to construct astronomical numbers of diverse antigen receptors if specificity can emerge from the population patterns generated by just a few receptors? I don't have a precise answer to your question, but I suspect that we shall find that the numbers of different T-cell clones we actually use are far less than the millions of clones we can generate potentially (§111). Be that as it may, the nervous system manages with just three color receptors because the brain supplies millions of neurons for the process of pattern formation. The immune system probably needs more antigen receptors than the eye needs color receptors because the immune system has no central brain to amplify its perceptions; the receptor-bearing lymphocytes alone have to create their own patterns of reactivity as they congregate *ad hoc* to react (§119). Also, the cones don't have to remember anything; the lymphocytes have to remember everything (§139).

§133 Context of Pattern

Above, we said that macrophages sense context and that the context of a signal is relative; what may be one's subject is another's context (§114). What is a context, and how is it relative? The concept of a pattern can help us functionally define the concept of *relative context*. The context of a signal is the *pattern* of other signals in which the designated signal is embedded. Macrophages recognize the arrangement of germ-line signals within which somatic antigen signals are embedded. The antigen signals, of course, are seen by the lymphocytes, not by the macrophages. A reactive pattern, in essence, makes no objective distinction between signal and context. The distinction between signal and context emerges from your point of view, depending on whether you happen to be a macrophage or a lymphocyte. We shall return to the concept of context when we discuss the immune language below (§137).

134 Stable Pattern Categories

The signals that jointly comprise a pattern can be said to belong to a common *category*. Thus IFNγ, TNFα, IL-1, and IL-12 belong to the T1 category of cytokines, and IL-4, IL-5, and IL-10 belong to the T2 category because their pleiotropic effects overlap in distinct patterns of immune reaction. Patterns operationally define categories. And categories provide tools for carving up reality into usable units (§57).

How stable are the patterns that create immune categories? It depends, like much of biology, on what, when and how much perturbs them. Sometimes a pattern can appear to be very robust and the deletion (or addition) of a key element may not seem to affect the system. Knocking out the IFNγ gene from a mouse genetically prone to autoimmune diabetes, for example, may not prevent the development of diabetes (§129). Interfering with IFNγ later in the course of life, however, can abort the diabetes pattern and cure the mouse. Note that categories and patterns are processes, not static compartments. Processes can sit securely in their basins of attraction (§22), and yet be very sensitive to specific control elements that may push the system into an alternative attractor basin, as we discussed above (§24). Treatment with the antibiotic to which the infection is sensitive can change the course of the disease; the wrong antibiotic is one to which the bacterium is not sensitive. The development of autoimmune diabetes can be sensitive or not to different cytokines depending on the state of the system. The key to control, thus, depends on discovering the control elements to which the disease pattern is sensitive.

COGNITIVE DECISION-MAKING

The immune system receives a pattern of signals, integrates them and responds by choosing a particular type of response pattern from the response repertoire. The decisions are made using an immune language that combines germ-line and somatic chemical words. Immune memory is founded on the

associations between somatic particulars and germ-line classes of behavior.
Immune choices emerge from the ongoing adjustment of molecular patterns
that reflect the ongoing needs of the body.

§135 Immune Decisions

In addition to self-organization and internal images, cognitive systems
are characterized by their ability to choose (§49). Above, we concluded
that cognitive choice emerges from the exercise of options (§50) influenced by history (§53). Choice, in other words, involves learning; the
residue of past experience is reflected in present behavior. Choice, as
defined here, is not dependent on any self-reflective consciousness or
mystically free will (§52). Like other cognitive activities, choice is deterministic.

The immune system can be said to choose because it both exercises
options and learns. The system has many optional patterns of seeing
and responding. The system can perceive almost any molecule as an
antigen, and it can deploy a large repertoire of potential responses. T1
and T2 cytokine patterns (§120) and the seven different types of antibody isotypes (§118) are just a few examples of the various effectors
that can take part in immune responses, responses which differ in the
way they activate genes and make cells grow, die, or move (§96).

The system is confronted with options, and so it must make functional
decisions each time it implements a particular course of action. Immune
decision-making is the process by which the immune system determines which response to append to which perceived signals. What
should I see and what should I do about it?

§136 Decision-making

Decisions arise from associations (§55). Immune effects are encoded in
germ-line reactions, while immune perceptions include antigens that are
encoded in somatic antigen-combining sites (§111). Decision-making,
therefore, often involves the task of associating somatic perceptions of
objects with classes of germ-line effector responses. If an immune
agent, a T cell for example, sees a peptide antigen on a target cell,

should the T cell kill the target cell or should the T cell stimulate the target cell to grow?

Immune decisions can also involve the association of germ-line signals with germ-line responses. If a macrophage becomes activated within a particular tissue, should the macrophage begin killing or should the macrophage begin healing?

Decisions associate images of perceptions with images of responses. Specificity, as we discussed above (§131), is embodied in patterns. Thus, the matching of perception patterns with response patterns is what we need to explain. How is it done? It's done rationally, by first talking it over.

137 Immune Dialogue and Decision-making

The decisions of the immune system are determined by the interactions of immune agents with their targets and with their fellow agents (co-respondence; §119). The character of this molecular dialogue can be analyzed by exploring five attributes of linguistic communication: abstraction, combinatorial signals, semantics, syntax, and context.

Abstraction. The immune system communicates by way of molecules, which are matter, but even molecules may serve as abstractions of another reality. For example, a processed peptide bound hand-and-foot in the cleft of an MHC molecule (P-MHC; §115) is a concrete entity, but it is not the actual virus or bacterium of which it was once a hidden part. At most, the peptide is a relic of an infectious agent. Indeed, in addition to serving as a representation of an infectious agent, a P-MHC compound ligand represents the state of the cell that presents the ligand. Cytokines and cell interaction molecules too can serve as abstract signals reporting broken bones or lacerated skin for immune system maintenance. The immune system recognizes not entities, but signs of entities. Just as a spoken word is both a physical reality and an abstraction, a molecule may function as a physical abstraction.

Combinatorial signals. Patterns are combinations of entities, and life fully exploits combinatorials. Combinatorials make it possible to use a limited number of elements to transmit a large number of signals. For example, an almost unlimited number of different proteins can be constructed from combinatorial strings of just twenty different amino acids, and an almost unlimited number of words from the 26 letters of the alphabet. Likewise, many combinatorial strings of different patterns can be made out of a few dozens of cytokines. For example, the missing IFNγ in the knock-out mice (§129) is apparently replaceable, at least in part, by combinatorial strings of various other cytokines. Such is the power of combinatorial strategies.

Semantics. The study of the relationship between a sign and its meaning is denoted by the term 'semantics'. The meaning of a sign, as we discussed (§8), can be defined operationally by the response the sign induces. Antigens, antibodies, cytokines, and other immune molecules can be said, therefore, to express semantic attributes because they induce measurable effects.

The semantics of natural language is a difficult subject; we understand a word when we hear it, but where does its meaning lie? The semantics of immune molecules are easier to explain than are the semantics of words. The semantic properties of immune agents can be reduced to their structure, defined at several levels: the conformations of ligands and their receptors, the arrangements of signal molecules in patterns, the repertoires of populations of cells, and so forth. All of these structures interact to produce meaning, functionally: nobody has to think about it.

Syntax. Syntax refers to the organized structure of a string of signals, the grammar of the message. At present, we don't know whether the order of immune signals is important in immune communication, but we can begin to classify the parts of immune speech, at least metaphorically. A processed peptide antigen seen by a T cell (the **P-MHC** ligand; §115) is like the subject (or object) of an immune response. The string of cytokines and other accessory signals presented along with the antigen tell the T cell how to respond to the antigen (T1 or T2, for example); these accessory signals thus *predicate* something about the antigen. In other words, an antigen can act like a 'noun', and the pattern of cytokines like a 'predicate' in a chemical 'sentence' spoken by a macrophage to a T cell. Note that antigens are defined somatically by the random creation of antigen receptors; the predicate signals, in contrast, are received in the germ-line. Immune grammar thus connects somatic and germ-line molecules in the same message. Immune grammar connects the individual's experience with the evolutionary experience of the species, across vast scales of time (§138).

Context. Patterns of signals create a signal context (§133). Immune molecules, like words, are much more meaningful when they are met within the context of a string of signals (a sentence) than when they are met in isolation; molecules out of context are intrinsically degenerate, redundant and pleiotropic (§131). The context in which a signal is embedded sharpens the meaning of the signal. The meaning of an antigen, the nature of the response it elicits, depends greatly on the cytokine context in which the antigen appears; a T1 mix of cytokines predicates a very different response to a given antigen than does a T2 mix of cytokines (§120). The context of the signal bears much of the meaning.

I would like to suggest that any system of social communication, whether of people or of cells, may be called a language if the transmitted messages incorporate an abstraction of reality, use combinatorial strings of signals, and combine semantics, syntax and context. If you agree with this assertion, I believe it reasonable to conclude that the immune system uses a language of sorts. I also believe that the immune language might share strategic structures with natural language worthy of study; they both are concerned with generating meaning out of information (§7, §8; also see H. Atlan and I. R. Cohen, 1998).

Of course you may argue that I am pushing the language metaphor too far, that the immune system does not use a natural language and that, consequently, immune communication can teach us nothing about natural language. I won't argue with you because it doesn't matter how you wish to define language. The process is more important than is the terminology. The important point is that immune communication proceeds as an ongoing chemical dialogue which functionally connects perception with response.

§138 Somatic and Germ-Line Connectivity

Immune connections between germ-line responses and somatic perceptions occur whenever lymphocytes, macrophages and tissue cells communicate using antigens expressed in the context of germ-line signals (§137). Such connections take place whenever a T cell recognizes a P-MHC compound ligand (§115) and co-responds with a B cell, and so helps the B cell secrete an antibody of a particular isotype (§118, §119). Somatic and germ-line connections are made whenever an antibody tags a target for destruction by a macrophage or NK cell (§114, §117). One could supply additional concrete examples of how the immune system connects the experience of the species to the experience of the individual. Immune interactions are an ongoing association of somatic particulars with germ-line classes of behavior. The associations can be reduced to defined molecular and cellular structures. Cognition emerges from the aggregate of these simple molecular interactions.

§139 Immune Memory

Memory is another cognitive concept whose mechanism is clearer in the immune system than it is in the brain. Memory is the expression of learning from past experience. Most of us can suffer measles or mumps only once in our lifetime because of immune memory. We may come down with a disease the first time we are infected with an infectious virus, because the virus flourishes while the immune system attempts to learn how to deal with it. We recover when the immune system musters a suitable response, one that associates the virus antigens with an effective germ-line response. Now we are immune (free of taxation; §91) to the virus. A second infection goes unnoticed; upon seeing the virus antigens, the immune system immediately mounts the response it has already discovered to be appropriate to destroy the virus.

What changes took place in our immune system after the first infection that now allow us to deal with the virus without having to pay the price of illness? There were two changes. First, the receptor repertoire of T cells, B cells and antibodies that can recognize the virus expanded. The second time around, the system is quantitatively better able to perceive the enemy. Secondly, and perhaps more importantly, the memory T and B cells were selected during the first infection to express the germ-line effector responses needed to kill the virus. The second response, from its outset, has the right quality.

During the first infection, the viral antigens were seen along with a variety of contextual signals. On the basis of the variant contextual signals, the responding T cells and B cells were activated to differentiate into T1, T2 and other response categories (§120). The successful responders were selected to become *memory* cells; having differentiated into a suitable effector type in response to the infection, the immune agents no longer require a full string of context signals to produce an effective memory response. Memory is the replacement of a context of infection by some antigens. Just as a name alone suffices to arouse associations born through the history of a relationship, the antigen alone now suffices to arouse the response suitable for the virus, without the disease. Immune memory is a more advanced state of cell differentiation.

Note, however, that memory is not an absolute or final state. Memories, as we all know, can be vivid or pale. Immune memory too comes in degrees. The memory response, like the primary response, still has to evolve. The immune dialogue still has to proceed. The second response is more efficient than is the first contact because the second response begins the dialogue from a more advanced state of self-organization. Even memory cells, however, modify their cytokine profiles as they respond and co-respond (§119).

Of course, you may never have to suffer either measles or mumps if you have been vaccinated with the attenuated viruses. Vaccination is a way to supply the immune system with the experience it needs to learn an effective response pattern to an agent of disease without incurring the actual disease. Vaccination usually does its job, but the classroom lesson is never as effective as is real experience on the street. The actual infection supplies more signals, teaches the system more comprehensively than any disease-free vaccine ever can. The most complete set of signals comes with the disease. That's why protection by way of vaccines usually needs boosting. But of course, health is more important than is completeness.

§140 Pattern Reflections

There is something paradoxical about immune activities; they seem to reiterate. The outputs are very much like the inputs; patterns of molecules go in and patterns of molecules come out. Moreover, the patterns in and out often contain the same molecules; tissue cells and immune cells mutually communicate by way of cytokines such as IL-1, IL-6, IL-8, IL-10, IL-11, TNFα, TGFβ, GM-CSF (never mind the designations), and by many of the same adhesion molecules.

It goes like this: the tissues activate the immune system by way of patterns of cytokines and cell interaction molecules, along with tissue antigens. The immune system, in turn, responds and co-responds (§119) by generating its own patterns of cytokines and cell interaction molecules, along with antigen receptors (§131). The body cells, in turn, respond to the pattern of immune signals and activate genes, grow, die, or move. And the immune cells too activate genes, grow, die, or move. The amounts of the molecules and their patterns produced by the body and by the immune cells differ, but in principle, the molecular patterns of the body are reflected in the molecular patterns of the immune system, and vice versa. In short, the interactions of the immune system with the body and with itself are expressed as molecular patterns reflecting other molecular patterns.

Note that the molecular patterns of the immune system and the body reflect each other dynamically; the patterns change and adjust. Like a dance, the mutual reflections evolve in time. The dance of molecular patterns between body and immune system begins with the development of the body and ends with the demise of the body. While there is life there is this reiterating, reflecting dance of molecular patterns. Immune activity is continuous. Reactive systems never leave the dance floor to rest on the sidelines.

If the immune system is in constant flux, how can it take the time to make cognitive decisions? Fortunately, the decision-making process doesn't need to take time out. Consider the way decisions are made on the field of play during a soccer match; the patterns of play constitute the decisions. Immune decision-making too is the dynamic self-organization of patterns of immune signal molecules, in constant interaction with signal molecules produced by the tissues. Patterns of immune response that evolve into a productive equilibrium within the body environment are favored, and patterns of response that fail to interact in a stable manner dissipate and disappear. The evolution of an immune response, like evolution in general, is the creation and occupancy of stable attractors (§30). Patterns evolve adaptively.

Reflection patterns might seem like a waste of time and energy; what can be gained from outputs that look like inputs? What do you gain from a dance of patterns? Why – you gain the action. Consider an economy: its input is money, resources, information and hard work, and its output is more money, resources, information and hard work. And the more activity there is, the healthier is the economy. Life too is constant interaction, reverberating interaction. Life emerges from molecular patterns reflecting a changing environment (§13), and decision-making is pattern-making to the end.

COMPARATIVE SUMMARY OF IMMUNITY

If you peruse a standard immunology text you will find little or no mention of many of the precincts we have visited on this tour of the immune system. Emergence, information, meaning, cognition, decisions, images, self-organization, degeneracy, co-respondence and patterns are not on the standard itinerary. Why not? The reason is simple: what you see depends on your viewpoint. Traditional immunology sees the immune system from the viewpoint of the clonal selection theory (CST). The first priority of the CST was to reduce the complex behavior of the immune system to the underlying chemistry of the interaction between an antigen and its antigen receptor. The classical CST assumes the specificity of the ligand–receptor interaction as a given fact; the CST emphasizes the adaptive protection of the individual from discrete dangers, and pays little attention to ongoing immune maintenance; and the CST makes no provision for regulating the response repertoire. The theory proposed by Gerald Edelman to describe the central nervous system is much closer to our description of the cognitive immune system.

141   The Clonal Selection Theory

During its formative years, modern immunology was reared by two parents: chemistry and microbiology. Ever since the discovery, about

100 years ago, that specific antibodies bind specific antigens, the main-line quest of immunology was to understand the molecular basis of antibody binding. The objectives were to uncover the chemistry of antibodies, to analyze how they are made and to learn how they recognize specific antigens. Protection against foreign invaders was thought to be guaranteed by the presence of antibodies specific for the invader. It was assumed by most, or at least hoped by most, that the complex biologic behavior of the system was inherent in the simple chemistry of antigen recognition.

The function of antigen recognition, in the eyes of the parental micro-biologists, was to protect the individual from foreign invaders, the causal agents of infectious diseases. The antibodies were thought to have evolved to mark infectious invaders for destruction by phagocytes or by effector molecules such as complement (§92). Specific immunity was in the hands of the lymphocytes that made the antibodies; the helper phagocytes and accessory molecules were the silent servants. Hence, it was felt that the behavior of the immune system could be boiled down to the chemistry of specific antibodies; complex biology would be reduced, by the scientific method, to fundamental chemistry (§5). The immune response, to borrow a metaphor from the nervous system, was like a reflex; the antigen was the stimulus and immune behavior was the automatic response. How is an antigen recognized as a stimulus? That was the question.

After some decades of probing for a unifying idea, the clonal selection theory (CST) of acquired immunity, put forth most effectively by Macfarlin Burnet in the late 1950s, was accepted as the standard paradigm (see S. H. Podalsky and A. I. Tauber, *The Generation of Diversity. Clonal Selection and the Rise of Molecular Immunology*, 1997). The CST taught that:

1. the lymphocytes, as a population, had to be born with a diversity of receptors ready made to accommodate all possible antigens (proved true and explained chemically some four decades later; §111);
2. each lymphocyte had to express an antigen receptor of only one specificity (essentially true, with the reservation that some T cells apparently can express two receptors);

3. any antigen entering the body will select those clones of lymphocytes that happen to have complementary ('specific') receptors and will activate them to secrete antibodies of the same specificity as the lymphocyte's antigen receptor (proved true, in principle).

If these three teachings of the CST are true, more or less, what is the problem? Why am I writing this book about a cognitive system? It's because the CST can no longer account for what we have learned about the immune system: the CST sees specificity where there is really degeneracy, it disregards ongoing body maintenance, and it neglects the problem of the response repertoire. We shall briefly outline here these anomalies of the CST; below, we shall discuss the views of the CST regarding autoimmunity (§147).

142 The Specificity of Clonal Selection

A major problem of immunology is to understand how the immune system avoids attacking the body itself. If, as predicted by the CST, the initial population of lymphocytes could potentially recognize all possible antigens, then each of us is born with autoimmune lymphocytes. Obviously, the immune systems of most people don't attack their owners. Burnet proposed that autoimmune lymphocytes might be born, but they could never be allowed to survive. An important corollary of the CST taught that the lymphocyte repertoire was purged early in development of any lymphocytes that happened to produce antigen receptors capable of recognizing self-molecules. Self-recognizing clones were forbidden, in the parlance of Burnet.

The problem of self-not-self recognition is an important issue in immunology, and we shall discuss the CST view of autoimmunity in greater detail below (§147). But note that the negation by the CST of self-recognition implies a fundamental assumption, one so basic that it seemed at the time to be unquestionable: the molecules of the self differ in their conformations (as antigens) from the foreign molecules of the outside world, and the antigen receptors of the immune system are capable of distinguishing the difference. Receptor specificity, in other words, is the starting point of the CST. The CST posits a one-to-one causal connection between the antigen and its receptor. Immune specificity, for the CST, is a given, and on that premise stands the CST.

Alas, receptors, we now know, are degenerate at the ultimate chemical and physical levels (§107). To reduce immune specificity to ligand–receptor chemistry is to kill immune specificity. The CST does not solve the specificity problem; it doesn't even recognize the existence of the problem. Immune specificity, as we discussed, is not a given; it is a construction (§112).

§143   The Transformations of Clonal Selection

A second shortcoming of the CST is that it neglects immune maintenance of the body (§96). True, defense traditionally has interested Western medicine more than has maintenance. People used to die dramatically from infections, and doctors were trained to save them acutely. People die only chronically from maintenance problems, and doctors are now at a loss. In any case, by neglecting immune maintenance, the CST distorts the basic character of the immune system.

If we assert that the immune system deals only with protecting the body against invasion, against the foreign, against danger, then we assume that the system is normally quiescent unless it is stimulated by antigens. The CST posits a sequential course of discrete events: the immune system is inactive until an antigen or some danger activates it; the immune response reflex generates immune effectors; the effectors reject the invader; the danger is gone; and the system returns to its resting baseline. Systems of such a character are called *transformational* systems, in the parlance of people in the information sciences who design computers and think about the essentials of systems (see Z. Manna and A. Pnueli, *The Temporal Logic of Reactive and Concurrent Systems*, 1992). Transformational systems transform information from one form into another in a defined sequence. They achieve their goal, and then they stop. They are *linear*.

Ongoing maintenance, as we have described the process, points to the immune system as a *reactive* system. Reactive systems, in contrast to transformational systems, are not sequential systems, they are *concurrent* systems; they never rest. Reactive systems are in constant dialogue with their environment, and with themselves. Reactive systems are ongoing, *non-linear* systems; they receive information and produce information in parallel in diverse ways. Reactive systems may include discrete transformational subsystems, which we can analyze in isolation. But the reactive system as a whole interacts dynamically with its environment and with itself concurrently. Reactive systems thrive on co-respondence (§119). They make maps and update them without stop. Decisions are never final. The brain is a reactive system; the economy is a reactive system. And my point here is that the immune system is a reactive system. Indeed, the format of the Figures we adopted beginning with Figure 17 (§91) suits the description of reactive systems. The CST, in neglecting immune maintenance, provides a simplistic and abridged picture of the immune system as a transformational system only. Times have changed, moreover, and society is now worried about chronic maintenance.

§144   The Incompleteness of Clonal Selection

Beyond its misunderstanding of specificity and its simplistic linear structure, the CST is no longer able to account for recent discoveries. The need for multiple recognition systems (macrophages, T cells, B cells), the need for co-respondence, the varied response repertoire, the world of cytokines, and the pleiotropisms and redundancies of the system have no place in the CST view of the world. Much research in the immunology is now devoted to understanding and controlling the response repertoire. The treatment of autoimmune disease, the development of vaccines against infectious agents and against cancer, and the control of transplantation all involve the response repertoire. But the CST provides no guidance in the matter.

§145   Neuronal Darwinism

We may not find the terms 'degeneracy', 'redundancy', 'co-respondence', or 'patterns' used in immunology texts. But such terms can be seen in neurobiology. Gerald M. Edelman has written a book *Neural Darwinism* (1987) in which he proposes a theory of neuronal organization that could account for the way the mammalian brain performs its cognitive functions. Edelman suggests that the brain dynamically maps the environment by deploying degeneracy and redundancy in multiple neuronal centers that independently record diverse features of the perceived world. The centers influence and update each other by a process, akin to our co-respondence, that Edelman calls *re-entry*. The brain self-organizes its networks of connections by a series of selections. Of course, neurons are not lymphocytes or macrophages; and, as we discussed, the two systems differ materially in many important ways (§95). Nevertheless, Edelman's theory provides support for the idea that the two systems defining our individuality may really use common operational strategies (§1, §2). A discussion of Edelman's theory is beyond our scope, and those interested should read Edelman.

Chapter 5
On Autoimmunity

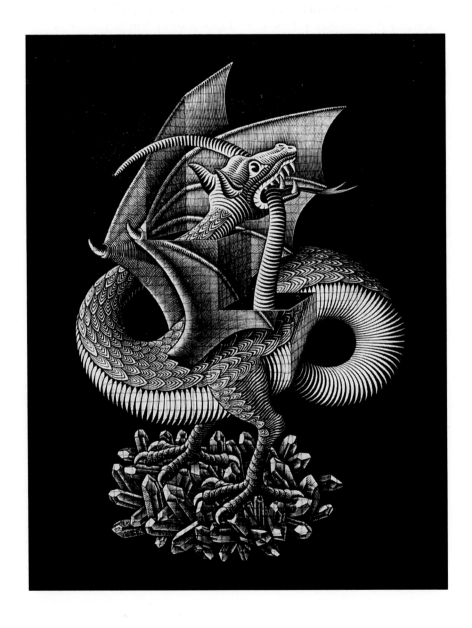

Chapter 5
On Autoimmunity

CLONAL SELECTION THEORY VIEWPOINT

Autoimmunity refers to the existence in one's immune system of antigen receptors that can recognize one's own molecules as self-antigens. The classical CST asserts that a healthy immune system must be purged of autoimmunity; autoimmunity cannot be physiological. Autoimmune diseases arise as random accidents.

146 Autoimmunity Defined

Autoimmunity is an awkward term, for both classical immunologists and for historians of words. The *auto* part of the term refers to the *self*, and there's no problem with that. The awkwardness is in the *immunity* which, as we saw, should mean freedom from penalty (§91). Autoimmunity, nevertheless, has come to mean the penalty a person could pay for housing an adaptive immune system. Autoimmunity describes a situation marked by T cells, B cells or antibodies whose antigen-combining sites can recognize self-molecules or parts of self-molecules. Obviously, all the systems of the body, the immune system included, recognize parts of the self; the body operates thanks to the interactions of germ-line receptors with self-molecules and other ligands (§13). But germ-line interactions are not considered autoimmunity. Autoimmunity refers to the capacity of one's somatically generated antigen receptors to recognize one's own molecules; by being so recognized, self-molecules become self-antigens (§111). Autoimmunity thus denotes the immune recognition of self-antigens.

We can imagine two types of autoimmunity: natural or physiological autoimmunity and autoimmune disease. An autoimmune disease may be defined as a pathologic condition caused by an actual immune attack directed against self-antigens. I emphasize *directed* against self-antigens. The body may be damaged by immune reactions directed against foreign antigens, as can happen if we are allergic to penicillin or if an immune reaction kills cells infected by a virus. But allergies and immune damage resulting from body protection are not autoimmune diseases. Defining words, however, is only a first step towards understanding. Understanding begins with questions. I shall structure our discussion of autoimmunity around five general questions.

Essence. What is the basis of autoimmunity? Is autoimmunity an essential part of the immune system, in the way that a tire is an essential part of an automobile? Or is autoimmunity incidental to the immune system in the way that a flat tire is incidental to the operation of the automobile? If incidental, is autoimmunity an unavoidable, though evil by-product of some good (like a well-worn tire that finally goes flat); or is it an unnecessary accident (like a blowout from a nail on the freeway)?

Organization. Is autoimmunity ordered or random; does the autoimmune receptor repertoire manifest any internal structuring of its components, any biases for particular self-antigens? If there be order, how does it come about?

Utility. Does physiological autoimmunity take part in any useful interactions; does it have a 'purpose'? Does autoimmune disease reflect any useful function?

Causation. What causes autoimmune disease? Are physiological autoimmunity and autoimmune disease related? Does the one arise from the other? If so, what accounts for the transition?

Therapy. How can we prevent an autoimmune disease? How might we cure a disease that has already arisen? What is the cost?

§147 The CST View of Autoimmunity

My description of autoimmunity is quite different from the classical view taught by the CST, but we shall begin with the CST view because it is simple to grasp and it is orthodox. The CST makes two points: first, the immune system of a healthy person must not and does not attack the person; hence, the immune system is *tolerant* of the self. And secondly, self-tolerance must arise from some mechanism that brings about 'the complete or partial elimination' of one's autoimmune clones (see M. Burnet, *Self and Not-Self*, 1969, page 230). In the zoology of the CST, physiological autoimmunity had to be a mythical beast.

The approach of the CST to autoimmunity arises, I believe, from the conviction that the biologic complexity of the immune response can be reduced to the one-to-one simplicity of a chemical reaction. The logic may be formulated like this: chemical reactions are regulated for the most part by the effective concentrations of the reactants. For example, molecules of oxygen and hydrogen will react to form molecules of water (at suitable concentrations, temperature, pressure, etc.) until one or the other reactant, the oxygen or the hydrogen, gets 'used up' and effectively disappears. Likewise, an immune response will be triggered and proceed as long as an antigen can be recognized by clones of lymphocytes bearing specific receptors for the antigen. In short, immune responses, like chemical reactions, should be regulated by the effective concentrations of the reacting partners.

The immune response of mature lymphocytes was seen by the CST to be like a reflex; if you see an antigen, attack it. Hence, any form of self-recognition would have to be expressed by an immune attack against the self, by a disease. The logic of the CST can be stated like this: if there is no autoimmune disease manifest in healthy individuals, there can be no autoimmune reaction, and if there is no reaction, then either the antigen is missing or the receptor-bearing clones are missing. But self-antigens abound; the body is made of self-antigens entirely. Hence, the absence of autoimmune disease must mean that there are no self-reactive clones in the healthy body. But such clones must arise, because antigen receptors are created randomly. Therefore, self-tolerance requires the elimination, during lymphocyte development, of any potentially self-reactive clones. By the same logic, the development of an autoimmune disease necessarily implies the accidental emergence of an autoimmune clone.

Burnet's theory of autoimmune disease and his negation of physiological autoimmunity make it easy to summarize the stand of classical CST regarding the five issues of autoimmunity:

1. *Essence.* Autoimmunity is a random accident, a blowout on the free-way.

2. *Organization.* There can be no order to autoimmunity. This is stated clearly by Burnet: 'It is of the essence of our approach to immunity that no two cases of autoimmune disease should be the same' (*ibid.*, page 257). Each disease is a chance accident, and any similarities between patients must be chance.

3. *Utility.* Autoimmunity can have no purpose. Autoimmunity is always forbidden.

4. *Causation.* Autoimmune diseases are caused by the failure to delete an autoimmune clone during its development, or by the mutation of the antigen receptor of a mature clone so that the clone now begins to recognize the self. There can be no connection between autoimmune disease and the healthy immune system.

5. *Therapy.* The only specific cure for an autoimmune disease is to kill or inactivate the forbidden autoimmune clone responsible for the disease.

The logic of the CST is impeccable, but an examination of the beast itself might lead one to quite different conclusions. Let us examine the five issues of autoimmunity without CST commitment. We shall begin with *organization* and leave the *essence* question for last.

ORGANIZATION OF AUTOIMMUNITY

Autoimmune diseases are well organized. There are few autoimmune diseases, and each expresses a stereotypical immune signature. Physiological autoimmunity exists, and it too is well ordered. Physiological autoimmunity focuses on a particular set of self-antigens forming a functional image of the self, the immunological homunculus.

§148 Autoimmune Diseases Are Ordered

Each of us expresses about 100,000 germ-line genes, which means that we express 100,000 different proteins, along with the additional types of molecules (sugars, fats, and others) that are synthesized by protein enzymes. Any of these self-molecules are potential self-antigens.

According to classical CST, any one of the 100,000 and more self-antigens could be the chance target of a forbidden clone. Hence, there should be more than 100,000 different autoimmune diseases in the population. Moreover, each case of autoimmune disease should result from an immune attack against one self-antigen. So the chances that any two people might suffer from the same autoimmune disease would be one in 100,000. Obviously, autoimmunity to different self-antigens expressed in the same body tissue, the liver for example, might produce the same clinical picture of liver damage. But, statistically, the underlying autoimmune reactions should be random and different in different patients. However logical, the facts about autoimmune disease contradict the predictions of the CST; autoimmune diseases are well ordered. The intrinsic order of autoimmune disease is manifest in three ways: there are a limited number of diseases, the various diseases are marked by a stereotypical autoimmunity to certain collectives of self-antigens, and patients show predispositions associated with their genes and gender. Autoimmune diseases come in discrete patterns.

Few diseases. It is beyond our scope to itemize the prevalent autoimmune diseases, and readers interested in clinical and immunological details should consult the medical literature. The relevant fact is that there are not 100,000 different diseases, not 1,000, and probably not even 100. Indeed, there are about two dozen or so medically defined diseases, and about ten or fewer can account for the vast majority of patients suffering from autoimmunity. Most people with an autoimmune disease suffer from:

- Multiple sclerosis.
- Type 1 diabetes.
- Rheumatoid arthritis.
- Vitiligo (patches of skin lacking pigment).
- Thyroid inflammation (Hashimoto's or Grave's diseases).
- Systemic lupus erythematosus (SLE).
- Inflammatory bowel disease (Crohn's disease or ulcerative colitis).
- Myasthenia gravis (muscle weakness).
- Liver inflammation (primary biliary cirrhosis or chronic active hepatitis).
- Destruction of blood platelets (idiopathic thrombocytopenic purpura; ITP).
- Destruction of red blood cells (hemolytic anemia).
- Eye inflammation (uveitis).
- Kidney inflammation (glomerulonephritis).

- Scleroderma (unnecessary scar tissue formed in the skin and other organs).
- Pemphigus (blistering of the skin).
- A few others.

Prevalent diseases like rheumatoid arthritis, Hashimoto's thyroiditis, and vitiligo can each affect one to two per cent of the population, if we count very mild forms. Type 1 diabetes and inflammatory bowel diseases have increased greatly in recent years in some developed countries, and the reader probably knows persons suffering from these diseases. Multiple sclerosis too is not uncommon. We shall discuss the increasing prevalence of autoimmune diseases below when we talk about causes (§162). Merely note that, in contradiction to the teachings of classical CST, autoimmune diseases are represented mainly by a few standard diseases and each disease has its own immune signature.

Standard autoantigens and collectives. The prevalent diseases manifest a notable uniformity in the self-antigens to which individual patients show heightened autoimmunity. Of the many proteins expressed in the thyroid gland, patients with Hashimoto's thyroiditis manifest autoimmunity to the protein thyroglobulin and to one or two other molecules. Many people with multiple sclerosis show autoimmunity to the same few brain proteins: myelin basic protein, MOG, or PLP (never mind the acronyms). Moreover, many autoimmune diseases are characterized by autoimmunity to a *collective* set of self-antigens. For example, patients suffering from SLE or developing type 1 diabetes share autoimmunity to characteristic sets of self-antigens, the sets differing, for the most part, in each disease. Indeed, the uniformities of autoimmune diseases can cross the boundaries between species. Laboratory mice bred to spontaneously develop SLE or autoimmune diabetes show the collectives of autoimmune reactions that characterize the spontaneous diseases in human patients. The visitation of an individual autoimmune disease is surely a madness, but there seems to be a method to its materialization.

Genetic predispositions. Patients suffering from particular autoimmune diseases share MHC genes (§116). In other words, one is more susceptible to developing a particular autoimmune disease if one has inherited certain MHC alleles. For example, most people who develop type 1 diabetes bear the MHC class II genes DR3, DR4 or DQ8; multiple sclerosis goes with DR2; Hashimoto's thyroiditis goes with DR5; and so forth.

Notice that I did not say that one will get type 1 diabetes, for example, if he or she has inherited DR3, DR4 and DQ8. Susceptibility genes are prerequisites, not causes; most people who bear predisposing genes will never suffer the autoimmune illness allowed by their genes. Monozygotic twins are individuals who have developed from a single fertilized egg cell and so have inherited the same germ-line genes (§1). Nevertheless, having a monozygotic twin with type 1 diabetes or with rheumatoid arthritis does not mean that you too will develop the disease; on the contrary, most twins are discordant (differ) when it comes to expressing an autoimmune disease. A particular MHC allele may be *necessary* for an autoimmune disease, but having the allele is not a *sufficient* cause for the disease. We shall discuss the necessary contribution of genes when we discuss the causes of autoimmune diseases below (§161). The point for now is that genetic predispositions imply that there is more intrinsic order to autoimmunity than there is to a blowout on the freeway.

Gender predispositions. The intrinsic order of autoimmune disease is also evident in the perplexing observation that there are marked differences between women and men in their susceptibilities to particular autoimmune diseases. SLE, for example, is ten-fold more prevalent in women than it is in men, and women predominate in Graves' disease of the thyroid (seven-fold), in rheumatoid arthritis and myasthenia gravis (three-fold), and in others. Female and male sex hormones have different effects on the immune systems of experimental animals. Females tend to resist infections better than males, and make more antibodies. It is possible that women and men have different immune systems because only women bear children. In any case, an increased tendency to autoimmune disease is built into the female immune system. The expression of autoimmune disease is no accident. What about the expression of physiological autoimmunity, autoimmunity without disease?

§149 Physiological Autoimmunity: The Immunological Homunculus

I have directed attention to the organization of physiological autoimmunity by giving it a name, the *immunological homunculus* (see I. R. Cohen, 1992; naming names is a first step in any new enterprise; see Genesis 2:19). *Homunculus* means *little man* (*homo,* 'man' in Latin). I borrowed the term from the *neurological homunculus,* which refers to the map of the body represented in the brain.

Neurologists have discovered that the cerebral cortex can be divided into distinct zones, each zone housing networks of nerve cells that are functionally related to particular parts of the body. The correspondence between body part and brain zone creates the brain's image of the body, a little man, an homunculus. A distinctive feature of the neurological homunculus is that there is no direct relationship between the size of the brain image and the size of the body part it maps. The amount of brain space devoted to a particular body part depends on the degree of control the brain must exercise in the function of the part. The human brain, for example, devotes a relatively large area to mapping the organs of speech, sight and facial expression, the dog's brain to the organ of smell, and the elephant's brain to the elephant's trunk. (Should we call the self-image in the dog's brain a 'caninunculus'; what would you like to call the elephant's homunculus?). Just as the brain maps the environment within which a creature must operate (§84), so does the brain map the individual's body that interacts with that environment. That functional map is the neurological homunculus.

I proposed that the immune system's image of the body be called the immunological homunculus. I was prompted to do so by the observation that physiological autoimmunity devotes more attention to some self-antigens than it does to others; certain self-antigens dominate physiological autoimmunity just as certain parts of the body dominate the brain map. Homuncular self-antigens are recognized with relatively high affinity by a relatively large number of autoimmune T and B cells. Your immune self-image, like your brain self-image, should serve your interactions with the world (§84). These suggestions, however, need to be proved; the term immunological homunculus is mainly a prescription for a research program.

The term homunculus, unfortunately, bears an historical connotation that could be misleading, and I want to dispel it now. In the period of the Renaissance in Europe, the mystery of development – how a fertilized egg becomes an adult body – was attributed to the prior existence in the sperm of a *little man*, an homunculus, who simply grew after fertilization into a big man (or woman). Don't scoff. The notion of a little man hidden in each sperm is not as primitive as it may seem; the idea of the homunculus was the embodiment of the idea of a master plan. Recall that DNA, in modern times, has been understood as a master plan (§41). Quite simply, the development of macroscopic complexity was felt to require the existence of a primary, underlying microscopic complexity. The 'little man' concept proposed that a complex system required a complex seed to begin with. Today, we know about genes, epi-genetics, and the self-organization of emergent complexity. We need no pre-existing 'little man' to explain the attainment of complexity either by the brain or by the immune system. The homunculus, for us, is merely a figure of speech, a shorthand designation for the images of the body that self-organize in the brain or immune system. History aside, which self-antigens are included in the immunological homunculus?

150 Homuncular Antigens

Physiological autoimmunity, both T-cell and B-cell, seems to be directed mainly to three types of homuncular self-antigens: immune molecules, maintenance molecules, and some tissue antigens.

Immune molecules. It may seem odd at first glance that the immune repertoire should contain antigen receptors that can recognize molecules of the immune system itself. However, physiological autoimmunity to immune molecules appears reasonable when we consider the concept of co-respondence (§119); autoimmunity to immune molecules can help the immune system respond to the states of activity of its own agents. The immunological homunculus includes autoimmunity to many immune molecules:

- *Antibody reaction sites*. The production of autoantibodies (anti-self antibodies) to the reaction sites (Fc domains; §118) of other antibodies appears whenever an intense immune response occurs. For historical reasons, such autoantibodies are called rheumatoid factors because they were first noted in patients with rheumatoid arthritis. But antibodies to antibody-reaction sites can be detected in the healthy immune response too. Note the reflexivity here: one antibody's antigen-combining site binds another antibody's reaction site.

- *Antigen-combining sites.* Autoimmunity to the antigen-combining sites of certain autoantibodies and autoimmune T cells is also included in the homunculus. Antibodies or T cells that recognize antigen-combining sites are termed anti-idiotypes because they recognize the unique identity, the idiotype, of their target antibody or T cell (§122). Anti-idiotypes provide an additional example of immune reflexivity: two antigen-combining sites bind one another – anti-autoimmunity, as it were. Anti-autoimmunity provides an effective way for autoimmunity to regulate itself, and we shall discuss this below (§153).

- *Cytokines, cytokine receptors, and complement* (§93). The immunological homunculus includes autoimmune reactivity to TNFα and other key cytokines, to cytokine receptors, and to effector molecules such as complement. Note here yet another sort of immune reflexivity: germ-line immune molecules serve as objects for recognition by somatically generated antigen receptors.

Maintenance molecules. To carry out its maintenance tasks, the immune system must be alert to the state of the body's cells. But how can the immune system detect the cells that need its attention? The expression of maintenance molecules is one important sign. Damaged cells unfailingly increase their expression of maintenance molecules when they deal with emergencies. Maintenance molecules, like p53 and stress proteins, not only maintain, they are the faithful signals of troubled cells (§64). Autoimmunity directed to maintenance molecules is metaphorically like the nose of the caninunculus that sniffs out danger; a cell's expression of maintenance molecules is metaphorically like a cell's cry for immune help.

Tissue molecules. Until recently, homuncular tissue antigens have been detected mainly by chance observation; a self-antigen noted to be a target in an autoimmune disease was later found to be the subject of physiological autoimmunity in healthy individuals. An example is the self-antigen myelin basic protein (MBP), a component of the myelin sheath that insulates the electrical signals in nerve fibers in the brain and spinal cord. It was observed that immunization of experimental animals to MBP could induce brain inflammation, damage to myelin, poor nerve conduction and paralysis, a condition now called experimental autoimmune encephalomyelitis (EAE). (EAE was occasionally induced in people in former days when rabies vaccines were inadvertently contaminated with MBP; fortunately modern rabies vaccines are free of this risk.) In addition to EAE, autoimmunity to MBP was detected in patients suffering from multiple sclerosis, suggesting that MBP may be a target in the human disease. However, autoimmune T cells reactive to MBP also can be detected in healthy humans and in experimental animals that have not been immunized to the antigen. The numbers of these autoimmune T cells and their degree of activation are lower in healthy individuals than they are in animals with EAE or in people with multiple sclerosis. But it is clear, nevertheless, that MBP is a major homuncular self-antigen, as well as a target of attack in autoimmune disease. Obviously, you would like to know the difference between harmless, physiological autoimmunity to MBP and the pernicious autoimmunity to MBP found in diseases like EAE and multiple sclerosis (§160).

151 Global Patterns of Autoantibodies

To grasp the dimensions of the immunological homunculus, we would like to be able to study physiological autoimmunity globally, not only the autoimmunity to this or that particular self-antigen, piecemeal – as was done, for example, with MBP. We would like to see the entire repertoire of physiological autoimmunity at a glance, note its intrinsic order and study the dynamic changes that might occur during different immune responses (§170). This may seem to be an unreasonable goal in view of the fact that immunologists traditionally have had to use purified antigens to detect specific immune agents one by one.

However, the recent work of the Portuguese immunologist Antonio Coutinho and his colleagues in Paris and Lisbon now makes possible the study of global patterns of homuncular autoantibodies. Rather than doing traditional immunology using defined self-antigens, the Coutinho group extracts the range of molecules expressed in various tissues – muscle, brain, skin, heart – and spreads out the extracted molecules on a suitable gradient, according to their relative sizes. The spread of self-molecules forms a matrix. The researcher then can test whether the blood sera of different individuals might contain auto-antibodies that bind to any part of the matrix. The test, as presently done, cannot identify which self-antigens are bound by the auto-antibodies, but each autoantibody can be marked by its binding 'address' on the matrix of self-molecules. The totality of autoantibod-ies bound to the matrix forms the homuncular pattern. The global antibody technique is in its infancy and needs considerable refinement. Nevertheless, several important observations have been made regard-ing physiological autoimmunity:

- The receptor specificities of physiological autoantibodies are directed to a limited number of self-antigens. Only about five to ten or so of the thousands of self-molecules extractable from each tis-sue appear to be targets for physiological autoantibodies. Moreover, the patterns of autoantibodies are fairly predictable; different indi-viduals manifest autoantibodies that bind to self-molecules with the same or similar matrix addresses. Global autoantibody patterns seem to form characteristic fingerprints for the individual and for the species.

- The global pattern of physiological autoantibodies does not appear to require experience with external antigens. The patterns are present from birth, and normal patterns can be seen in experimental animals that have been maintained free of contact with infections or foreign antigens. The contact of the immune system with the body itself suffices to induce the normal pattern of autoantibodies.

- Some autoantibodies seem to be masked by anti-idiotypic anti-bodies (anti-autoimmunity; §153). Many natural autoantibodies become detectable only after their matching anti-idiotypic antibodies are removed.

- Immune responses to foreign antigens can transiently modify the patterns of natural autoantibodies. The homuncular pattern of autoantibodies, however, returns to normal as the reaction to the foreign antigen wanes. Physiological autoimmunity reflects, there-fore, the global state of the immune system. Autoimmunity accompanies immune activity generally.

- The development of an autoimmune disease can be associated with marked changes in the normal global autoantibody pattern. In contrast to the transient shifts in pattern that are seen in a healthy immune response, the abnormal pattern that accompanies an autoimmune disease may persist. An autoimmune disease, therefore, is not an aberration of a single autoimmune clone as taught by the CST (§147); an autoimmune disease reflects far-reaching changes in the whole immune system.

Antonio Coutinho and his colleagues did not design their global analysis technique to study the immunological homunculus, but their observations are convincing evidence for the reality of the homunculus concept. Investigation of the immunological homunculus will be aided by improvements in the technology and analysis of global antibody patterns, and by the identification of the individual self-antigens of the homuncular set. We also need to develop a way to study the entire set of T-cell homuncular antigens. Nevertheless, it is clear that our immune system possesses a dynamic picture of ourselves.

§152 Origin of the Immunological Homunculus

The T-cell repertoire is formed by the positive selection of newborn T cells that bear antigen receptors moderately degenerate for self-P-MHC ligands (§125). But if all T cells are degenerately autoimmune, how does the immunological homunculus come to focus its T-cell attention on a particular set of self-antigens? The answer is not yet known. Antigen receptors are generated at random (§111), so a bias for homuncular self-antigens cannot be intrinsic to the unselected repertoire. It is more reasonable to suspect that the homuncular self-antigens themselves are the prime movers in forming the homunculus. One can imagine that the homuncular antigens are programmed (in the germline) to be expressed in states, times and places where they can select autoimmune T cells with a sufficiently high affinity to drive them into clonal expansion. In support of this idea, MBP and other homuncular self-antigens have been found to be expressed in the thymus, the site of T-cell maturation and primary selection.

Thus, it is conceivable that the thymus presents self-antigens in two ways: non-homuncular and homuncular. Non-homuncular self-P-MHC ligands are expressed on the pathway that leads to the selection of degenerate clones that do not immediately expand their numbers of offspring. These T-cell clones may proliferate and undergo additional selections later, if they chance to meet altered, foreign anti-gens in the periphery (§126, §127). Selection for high affinity and early clonal expansion is reserved for homuncular self-antigens. This hypothesis, if correct, can provide another example of how the germ-line experience of the species molds individual somatic experience (§138).

§153   Homuncular Anti-Autoimmunity

The immunological homunculus appears to be regulated by anti–idio-typic T cells and B cells that recognize the receptors of homuncular autoimmune lymphocytes. Thus the immunological homunculus is composed of three different elements: a set of particular self-antigens, autoimmune T cells and B cells that recognize this set of self-antigens, and anti-idiotypic T cells and B cells that recognize, in turn, the autoimmune cells themselves. To put it more succinctly, the homuncu-lus is made of self-antigens, autoimmunity, and anti-autoimmunity. The clones of anti-autoimmune T cells and B cells, as we shall discuss later, can help restrain physiological autoimmunity so that, even when activated, physiological autoimmunity does not erupt into autoimmune disease (§168).

The origin of the anti-autoimmune regulator cells is presently obscure, but it is conceivable that the expansion of homuncular autoimmune cells in the thymus (or bone marrow) might lead to the selection and expansion of *anti-autoimmune* lymphocytes that recognize the autoim-mune antigen receptors. There is some evidence that activated T cells can process peptides of their antigen receptor and present them as part of an MHC compound ligand to other T cells. Thus the expression of homuncular self-antigens, encoded in the germ-line, is proposed to induce complementary autoimmune lymphocytes, which, in their turn, induce anti-autoimmune (anti-idiotypic) lymphocytes. The homuncu-lus is a reflexive troika. Figure 31 summarizes the development of the immunological homunculus.

Newborn T cells are positively selected in the thymus along two path-ways. One selection is for low affinity recognition of *non–homuncular* self-peptides associated with MHC molecules (**P-MHC**; §125). These T cells are destined to provide *anti-foreign immunity* by responding in

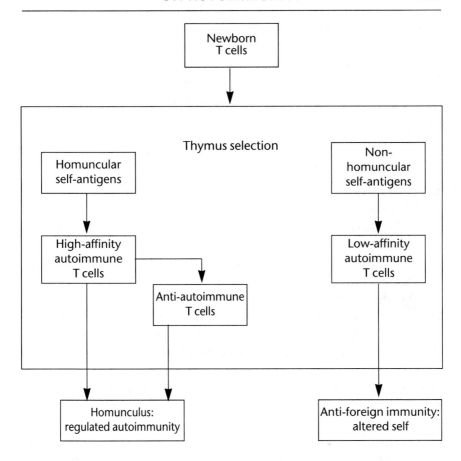

Figure 31: The development of the immunological homunculus

the periphery to the processed peptides of foreign antigens that fit the T-cell receptors with high affinity – a type of *altered self* (§126). The second pathway of T-cell selection appears to be made by *homuncular self-antigens*, which select for *high affinity autoimmune T cells*. In some way, these T cells are able to select *anti-autoimmune T cells*. The auto-immune and anti-autoimmune T cells regulate homuncular autoimmunity physiologically.

UTILITY OF AUTOIMMUNITY

Autoimmunity to homuncular self-antigens serves immunity in two ways. First, it modulates immune responses generally, both in turning on and turning off reactivity to other antigens. Secondly, it specifically participates in tissue maintenance, rejection of foreign tissues, protection against tumor cells, and enhancement of immunity to infectious agents.

§154 Homuncular Co-respondence

Physiological autoimmunity to homuncular self-antigens can influence immunity generally, paradoxically both up-regulating and down-regulating immune reactions to other antigens. The homunculus co-responds with other immune responses and up-regulates reactivity because of the high frequency of its autoimmune cells. The homunculus also can down-regulate reactivity because of its built-in anti-autoimmune regulatory mechanisms.

Homuncular up-regulation goes like this. We are born with a low frequency of T cells (and probably with a low frequency of B cells) to any given foreign antigen; any foreign antigen would have to appear structurally as a high-affinity alteration of a standard self-P-MHC ligand (§126). Thus, a primary immune response to a foreign antigen begins slowly and proceeds slowly as the rare T-cell clones begin to recognize the antigen, proliferate and differentiate to produce immune effects. The dynamics of a primary reaction can be intensified, however, by a conducive environment of cytokines. And that stimulatory environment can be supplied by the homunculus.

Now, important immune responses are marked by the increased expression of homuncular self-antigens because important immune responses are accompanied by tissue damage and cellular stress. So the homuncular autoimmune T cells, which are present in high frequency, are readily activated at sites of emergency, and these T cells secrete cytokines. These cytokines predicate, as it were (§137), the accelerated reaction of the few T cells recognizing the foreign antigens. The homunculus thus acts as an internal adjuvant that promotes immune reactivity generally.

Homuncular down-regulation goes like this. A primary immune reaction to a rare foreign antigen has no built-in regulation; thus a response to a foreign antigen may be difficult to terminate before the last trace of foreign antigen is destroyed and removed. Some antigens are not easily removed; how can the reaction be shut off despite the persistence of antigen? Fortunately, homuncular autoimmunity includes anti-autoimmune regulators that act to control the process (§153). Once the damage is reduced, the anti-autoimmune regulators secrete anti-inflammatory cytokines that provide the suppressive environment needed to terminate the ongoing response to residual foreign antigens. The immunological homunculus thus allows the state of the tissues to influence the general degree of immune activity in the neighborhood. Homuncular autoimmunity is a form of adjuvant co-respondence (§119).

§155 Homuncular Maintenance

Tissue regeneration, healing, scar formation, and blood vessel development are regulated to various degrees by the activities of immune molecules (§96). Homuncular T cells and B cells can aid body maintenance by secreting healing cytokines and growth-promoting molecules at sites of damage. The homuncular self-antigen MBP (§150), for example, is situated on the inner face of the myelin sheath, and is largely hidden from immune cells and antibodies. Intact MBP is exposed to the immune system following damage to myelin, and the exposure appears to activate the natural anti-MBP autoimmunity present from birth. In fact, there is reason to believe that activation of homuncular autoimmunity to MBP may actually help preserve the central nervous system after traumatic injury (see G. Moalem, R. Leibowitz-Amit, E. Yoles, F. Mor, I. R. Cohen and M. Schwartz, 1999).

Damage to other tissues too leads to activation of homuncular autoimmunity. A heart attack, for example, is marked by the transient activation of autoimmunity to cardiac myosin, a molecule normally sequestered inside intact heart muscle cells. Autoimmunity to stress proteins is also transiently evident at sites of tissue inflammation of any type. Thus, homuncular self-antigens, exposed as a consequence of damage, mobilize beneficial autoimmune maintenance to the sites where it is needed. Physiological autoimmunity is not merely innocuous; it serves to maintain the body. Immune maintenance, thanks to the homunculus, does not require the presence of foreign antigens. But know that this service, like all services, costs; nothing in this world is for free (§9). Imagine what happens when maintenance autoimmunity fails to get turned off by anti-autoimmune regulation; you might guess that an autoimmune disease emerges. More about that later (§160).

§156 Autoimmunity and Transplantation Immunity

Transplantation immunity refers to the attack of one person's immune system against the tissues of another person. You might want to applaud transplantation immunity as a safeguard to your individuality. But transplantation immunity is the major barrier to the replacement of worn-out or diseased organs – kidneys, hearts, livers. It is also the major barrier to the displacement of cancerous blood cells and other abnormal stem cells.

There are two general types of transplantation reactions: host-versus-graft (HVG) and graft-versus-host (GVH). The HVG reaction refers to the reaction of a person's immune system against tissue transplanted from another person – called an allograft (*allo*, 'other'). The GVH reaction refers to the immune attack suffered by an individual who has received immune cells from another person; in this case the immune cells of the graft attack the graft recipient, the host. Clinically, the GVH reaction prevents the transplantation of bone marrow to people in need; the immune cells in the bone marrow, in the absence of treatment, can attack and kill the recipient. Both the HVG and the GVH reactions are due to the intolerance of the immune system, mostly of the T cells, to foreign tissues.

Transplantation reactions against tissues are marked by characteristics that seem to set them apart from standard immune reactions to foreign antigens.

- The inciting trigger of a transplantation reaction is the foreign MHC molecules of the graft (§115). A one-molecule MHC difference suffices to trigger rejection; a transplanted organ that is otherwise genetically identical to the recipient, but which differs by only one MHC molecule, can be rejected as viciously as is an allograft that is foreign throughout. By the same token, an allograft that is identical to the host in its MHC molecules, but different in all other molecules, can be tolerated for long periods without being rejected. (Indeed, you now know how the MHC got its name; the <u>M</u>ajor <u>H</u>istocompatibility <u>C</u>omplex – the chief factor in the compatibility of grafted tissue; *histo*, 'tissue'.)

- The primary selection of T cells in the thymus seems to suffice for the transplantation reaction; the allograft rejection reaction is quite vigorous even in the absence of a second selection (§126). A primary transplantation reaction can be induced in the test tube merely by mixing allogeneic immune cells. A primary immune response to non-MHC antigens usually cannot be induced in a test tube. The primary response to a foreign MHC graft acts like it were a memory response.

- The numbers of T cells that seem to be able to recognize an allograft, even an allograft with a single MHC difference, are enormous: perhaps 5 per cent of all the T cells. How can a single foreign molecule activate 5 per cent of your T-cell clones? If your T cells can distinguish between different MHC alleles, and they can, then simple arithmetic says that twenty different MHC alleles will exhaust your T-cell repertoire. How can the repertoire see the many thousands of different antigens that we know it can, and does see? Unless foreign MHC molecules are not simple antigens; and they are not.

What is there about foreign MHC molecules that endows them with their special powers? They turn on the immunological homunculus, all at once. The story is like this. True love, we are told, does not alter its response when it alteration finds (see W. Shakespeare, Sonnet 116); but T cells do alter their response, and strongly when they meet altered antigens. A T cell is first selected in the thymus to recognize a self-**P**-MHC ligand with degenerate affinity (§125). The second selection of the T cell occurs in the periphery when its receptor binds an altered **P**-MHC ligand that fits it with higher affinity than did the ligand which selected it in the thymus. In other words, a foreign antigen activates a strong T-cell response when it looks like a self-antigen altered to bind the T-cell antigen receptor with high affinity (§126). T cells seek alterations. (I suppose some lovers do too; see W. Shakespeare, Sonnet 93.) Think of maintenance as the need to deal with the altered self.

Thus the transplantation reaction is not triggered directly by foreign MHC molecules; the reaction is triggered by the totality of peptides that can be presented as altered ligands (super-agonists; §108) by the MHC molecules of the graft. That's why so many T-cell clones are involved. The allograft reaction is an anti-altered-self reaction – autoimmunity in extremis. The question, of course, is whether the anti-altered self response to the allograft is especially triggered by altered *homuncular* self-antigens. The strength of the transplantation reaction and its dynamics would suggest that the homunculus is very much involved; but more work needs to be done to answer the question. In any case, know that the other is an altered you.

Indeed, MHC molecules that are too far different from those of the individual's own MHC molecules don't activate a strong T-cell mediated transplantation reaction. For example, grafts form another species (xenografts; *xeno*, 'foreign' in Greek) such as from a pig are much less strongly attacked by T cells than are allografts from other humans. Xenografts won't solve the rejection problem, however; xenografts are strangled by existing host antibodies that acutely cut off the graft's blood supply: but that is a different issue.

§157 Homuncular Destruction of Tumor Cells

Immunosurveillance is a term for the idea that the immune system is responsible for detecting newly developing cancer cells and killing them before they might proliferate to become tumors. The immuno-surveillance idea arose some decades ago in the wake of the observation that experimental animals might sometimes be successfully immunized against transplanted tumor cells. The idea of immunosurveillance was controversial; some found it attractive, while others did not, the issue being whether one believed tumor cells did or did not express unique antigens. According to the teachings of the CST, the immune system would have no way of detecting a potential tumor cell unless that cell was marked by a tumor-specific (not-self) antigen. Discussion of immunosurveillance eventually fell out of fashion because the issue of tumor antigens could not be resolved. There was no way of studying successful surveillance because a successfully killed cancer cell could not provide us with its antigens for investigation; likewise, there was no fair way to study surveillance against actual tumors because such tumors, in order to grow, had to have already evaded surveillance (by not showing their antigens, for example). The immunological homunculus, of course, advocates a change in our basic thinking about immunosurveillance.

The reality of physiological autoimmunity allows one to imagine a variety of ways the immune system might reject a potential tumor cell, irrespective of whether or not the tumor cell expresses unique antigens. Homuncular self-antigens could serve as targets of immune attack under a variety of circumstances: tumor cells, because of their aberrant genes and growth patterns, express maintenance molecules like p53 and stress proteins (§98) at very high levels, which could activate natural autoimmune agents to attack (§150); tumor cells can express altered self-P-MHC ligands or predicate molecules (cytokines, adhesion molecules) that provide an inflammatory context of immune dialogue (§137); tumor cells, by their abnormal nature, are prone to express abnormal patterns of molecules of immune interest (§140). Thus there are many valid reasons to expect immunosurveillance to work, even without having to invoke elusive tumor-specific antigens. In fact, we now know that germ-line mechanisms in macrophages (§114) and NK cells (§117) can kill tumor cells without help from somatic antigen receptors. Considering the great investment of life in apoptosis (§97) and the maintenance functions of the immune system (98), it would be very surprising indeed if immunosurveillance did not exist. With these possibilities in mind, new experimental investigations of immuno-surveillance might be undertaken. Certainly, the immunological homunculus offers some ideas for the design of tumor vaccines to pre-vent tumors, or to induce the immune rejection of tumors that have arisen. Think of anti-tumor immunity as an autoimmune 'disease' in which the target tissue is the tumor.

§158   Homuncular Protection Against Infection

According to the teachings of the CST, the more an antigen is similar to a self-antigen, the fewer the clones of lymphocytes there could be that might recognize it (§147). The CST would predict, therefore, that the immune response to an infectious agent is likely to be directed primarily to its most immunologically foreign molecules. This, how-ever, does not seem to be the case. On the contrary, your response to a bacterium or virus is often directed to its antigens that are most like your own. In the immune response to the tuberculosis organism, for example, about 20 per cent of the responding T cells seem to be specific for only one of the organism's thousands of potential antigens: a stress protein called hsp60, which is identical to your own hsp60 molecule in about 50 per cent of its amino acid sequence. This focus of the immune system on a self-like antigen is not unique to tuberculosis. The immune response to many different infectious agents seems to prefer molecules that, like hsp60, are highly conserved in evolution and very similar to self-molecules.

Molecules that solve universal problems for living organisms tend to be conserved from one creature to the next throughout evolution (§28). Evolution is modular; a newly evolved organism need not, and cannot, reinvent all its wheels. Creatures with markedly different lifestyles are obliged to use at least some common molecules. Maintenance molecules are essential for all cells, and so they are shared by many different forms of life. For this reason, your physiological autoimmunity to stress proteins and other essential molecules prepares you for your encounter with the essential molecules of your potential parasites. The similarities between your maintenance molecules and theirs makes their maintenance molecules look like alterations of your own. And, as we know, the immune system is very sensitive to altered ligands (§156). Hence, parts of the immunological homunculus become activated in the course of immune responses to infectious agents. The activated autoimmune cells produce cytokines that create a context of inflammation that enhances the sensitivity of the immune response to the foreign and partially foreign antigens of the infectious agent. Thus, the immunological homunculus helps prepare the individual for his or her lifelong struggle with infectious parasites. Natural autoimmunity has its advantages; by learning to know yourself, you learn to know your enemies. Figure 32 schematically illustrates the utility of homuncular autoimmunity.

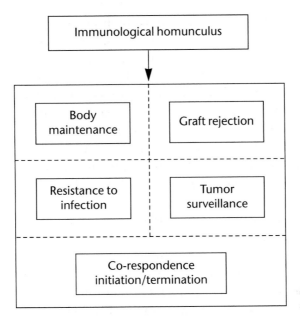

Figure 32: The physiological functions of the immunological homunculus

CAUSATION OF AUTOIMMUNE DISEASE

Autoimmune diseases emerge from dysregulation of the immunological homunculus. Physiological autoimmunity has to be activated as the need arises to maintain the body. Inappropriate persistence of healing as well as tissue destruction leads to disease. An autoimmune disease is an undesirable attractor, a damaging pattern of reactions. Autoimmune malfunctions are caused by conflicting signals influenced by genetic predispositions, immune responses to infectious agents, and tissue susceptibilities.

§159 Back to Causes

Science, as we discussed, seeks to reduce complex phenomena to one or more of the five sorts of causal explanations we outlined earlier: a fundamental law of nature, an agent–event relationship, a structure–function relationship, an energy transaction, or a transfer of information (§6). To blame any of the five factors for causing a disease, we would have to show the factor to be both necessary and sufficient to produce the disease; the factor must be present to develop the disease, and its presence must make the disease inevitable. The CST proposes that the presence of an autoimmune clone is both necessary and sufficient to cause an autoimmune disease; the CST would like to reduce autoimmune disease to an agent–event relationship. Defective immune *agents* (§91) are the cause, according to the CST.

The CST view, however, has no explanation for the regularity of autoimmune disease, nor can it explain the existence of physiological self-recognition without disease. Most importantly, we cannot attribute the development of a spontaneous autoimmune disease to any single necessary and sufficient cause. I want to make the point that autoimmune diseases are not reducible to one or another defective agent working in isolation; rather, autoimmune diseases emerge from the interactions of many factors (§12). The blame is not on the agents, but on their interactions in time and in space (§91). The interactions that give rise to the disease are not unlike those that occur in the activation of physiological autoimmunity, except that the pathological interactions persist as an attractor (§19). A pathological attractor is a pathological pattern of actions and reactions (§131). The injurious interactions can be related, I believe, to malfunction of the immunological homunculus through faulty interpretation of conflicting signals.

§160   The Misguided Homunculus and Autoimmune Disease

Look at it this way. To maintain the body, physiological autoimmunity needs to exercise power; we know it can reject altered cells, enhance immunity to infectious agents and heal damaged tissues. We would like to hope that the homunculus has the power to destroy tumor cells. In each of these situations, physiological autoimmunity is activated to perform various effector reactions. And, as we know, whatever in this world has the power to help also has the power to harm.

Disease can develop from the unnecessary destruction of body tissues; that is clear and easy to understand. But an autoimmune disease also can arise from unnecessary and persistent processes of healing. Joints are destroyed in rheumatoid arthritis, for example, by unbridled healing, from the invasion of scar tissue (called pannus) into the joint cartilage and bone. The filters of the kidneys too can become clogged and damaged by unregulated scar formation. Scleroderma results from abnormal scar tissue formed in the skin. Inappropriate formation of blood vessels can destroy delicate structures in the eye.

Autoimmune diseases are not caused by the mere activation of autoimmune clones, which occurs from time to time physiologically. Autoimmune diseases are caused by the failure of autoimmunity to become deactivated. Activation of physiological autoimmunity that becomes unsuitably recurrent or chronic marks the transition from physiology to disease. Disease is autoimmunity stuck in a rut, an undesirable attractor (§19). Inappropriate destruction and healing result from a faulty dialogue between the immune system and the tissues. Autoimmune diseases arise from conflicting signals (§163), poor decisions (§137) and feeble anti-autoimmune regulation (§153). Homuncular malfunctions are influenced by genetic predispositions, infectious agents, and tissue susceptibilities. Let us examine each interacting factor.

§161 Genes in Autoimmune Disease

Genes and how they work can be easily studied in experimental animal models; animal models allow us to do genetic experiments that we cannot do in humans. Inbred animals are especially useful. Strains of animals are designated inbred, or pure bred, when each animal in the strain is genetically identical in its germ-line inheritance (except that the males have one Y chromosome and one X chromosome, and the females two X chromosomes and no Y chromosome). Inbred animals are even more homogeneous than are identical twins because the copies of DNA inherited from mother and father are the same, except for the male Y chromosome. Having an inbred strain of rodent is equivalent to having an unlimited number of copies of the same germ-line rodent. So it is possible to use inbred rodents to carry out precisely controlled and repeatable experiments; each experiment can be done multiple times and in different ways, as it were, on standard copies of the same creature. Of late, we can even construct mice with particular genes removed (knocked out) or added (§129). Of course, each mouse or rat, like the hypothetically cloned Michael Jordans, will have its own brain and its own immune system (§1, §2).

Particular strains of rodents have been very helpful in elucidating important autoimmune genes. Mice of the inbred strain called NOD (for <u>N</u>on-<u>O</u>bese <u>D</u>iabetic) spontaneously develop autoimmune diabetes (called type 1, or juvenile diabetes) that is very similar in its clinical expression and immunology to human autoimmune diabetes.

(You might infer from the non-obese diabetic that there is another type of diabetes marked by obesity, and you would be correct; the obese type of diabetes – called type 2 – is not an autoimmune disease in either mice or humans.)

Through breeding experiments, NOD mice have been found to express at least fifteen genes that contribute to their susceptibility to diabetes. The unique MHC-II gene of the NOD mouse is critical, as you would expect. A human MHC-II gene associated with susceptibility to diabetes called DQ8 seems to be very similar to the NOD mouse MHC-II gene in the peptides (**P**) both can present to T cells (§115). Thus the mouse and human MHC-II genes may encode similar homuncular images relevant to the development of diabetes. Your homuncular picture of yourself is critical; autoimmune diabetes, SLE, rheumatoid arthritis, multiple sclerosis and the others, each requires a certain pattern of physiological autoimmunity as a foundation for its pathology.

The other fourteen and more genes involved in diabetes in the NOD mouse would seem to influence the activation of the immune response and its regulation. Gender is also important; female NOD mice show a much higher incidence of diabetes than do the males (§148). I would guess that equally large numbers of different genes will be shown to have some effect on any autoimmune disease.

The fact that many genes contribute to susceptibility indicates that no single gene is a sufficient cause; as we know, too many excuses mean that no one of them is a sufficiently good excuse. Each gene adds or detracts some element from a complex interaction, an attractor (§19). True, autoimmune diseases are rooted in a permissible genetic soil. But autoimmune diseases are not genetic in the sense that a disease like sickle cell anemia is genetic. Sickle cell anemia appears whenever a person has inherited from both mother and father a certain abnormal hemoglobin (the protein that carries oxygen in red blood cells), and that's that. The gene for the abnormal hemoglobin molecule is both necessary and sufficient to develop the anemia. The anemia can be reduced to the abnormal structure of a single protein, to a basic structure–function relationship (§6). Thus far, however, we know of no single immune system gene that is both necessary and sufficient for an autoimmune disease. So please don't be concerned about your genes; your chances of not getting an autoimmune disease outweigh your susceptibility, no matter what your genes may be. Moreover, the genes that encode susceptibility to one autoimmune disease probably encode resistance to other diseases. Autoimmunity is a trade-off. An inbred NOD mouse with fifteen genes all pushing in the direction of autoimmune diabetes can avoid diabetes if the mouse has once suffered an infection, as we shall discuss below. The immune system self-organizes beyond the genes, and experience is formative (§129).

162 Infections Can Prevent Autoimmune Disease

An immune response to an infection, as we mentioned above, normally activates the homunculus (§151). This activation is useful; it can actually help you fight off the infection (§158). But the activation of natural autoimmunity by the infection can also modify your susceptibility to an autoimmune disease. Paradoxically, the immune response to an infectious agent can induce an autoimmune disease, or prevent an autoimmune disease, depending on the circumstances. Here is an example of prevention.

Infection plays a decisive role in the diabetes of NOD mice; infection-free female NOD mice develop a very high incidence of diabetes, up to 95 per cent, while female NOD mice exposed to infections may have an incidence of diabetes of only 5–10 per cent. Thus, the immune system of an NOD mouse, by practicing on infectious agents, can learn to become more adept at controlling its genetic tendency towards autoimmune diabetes. Anti-autoimmune regulation is strengthened by self-organization through experience with infectious agents. Knowing others helps you know yourself.

Could a lack of immune experience with infections also make humans more susceptible to diabetes? We don't know. But consider the fact that the incidence of autoimmune diabetes has increased greatly in the past 50 years, and continues to increase in the developed countries of the

Northern Hemisphere. Has the antiseptic environment of the well-to-do family with its few, well-protected children led to immune deprivation? The immune system, like the brain, needs experience to self-organize (§2). Evolution has counted on recurrent infections for epi-genetic immune self-organization. Perhaps we, who can afford it, are depriving our children of needed immune experience (see G. A. W. Rook and J. L. Stanford, 1998). True, a decrease in childhood infections is not the only recent change in the environment of the well-to-do; the environment is polluted with novel chemicals. But the NOD mouse makes us wonder about the price we may be paying for the eradication from our niche of problems we have evolved to live with. Perhaps evolution, like the talented person, has learned not only to live with its problems, but to thrive on them.

§163 Infection, Conflicting Signals and Autoimmune Disease

In contrast to the story of the NOD mouse, immune responses to some infectious agents have been shown actually to induce autoimmune diseases. A clear example in humans is acute rheumatic fever, in which a throat (or skin) infection with a certain type of Streptococcus bacterium can induce autoimmune inflammations of the joints (arthritis), heart (carditis), or kidneys (nephritis). Reactive arthritis is a form of arthritis that can follow infection of the gut with a variety of bacteria. Inflammatory bowel disease (Crohn's disease) is another autoimmune problem that seems to be activated, or at least aggravated, in susceptible subjects by colonization of the gut by bacteria. Note that the bacterial culprits in inflammatory bowel disease seem to include bacteria normally found in the bowels of healthy subjects, bacteria that rarely if ever penetrate the tissues from their niche in the gut. In each of these examples, the autoimmune disease is caused by the host's immune response to the infectious agent, not by any intrinsic toxic property of the infectious agent. How can an immune response to a foreign entity trigger a damaging autoimmune reaction to the self?

Before answering, let me tell you about a concoction called by immunologists Complete Freund's Adjuvant (CFA). Jules Freund, an immunologist who worked in New York in the middle years of the twentieth century, discovered that especially strong immune responses could be induced to almost any antigen by immunizing the subject with the antigen emulsified in mineral oil containing killed Mycobacteria organisms. Mycobacteria of certain types can cause tuberculosis infections, but not, of course, when the Mycobacteria are dead. Immunologists, with their attention focused on the immune response to the added antigen, viewed the mixture of oil and Mycobacteria as an additive, an *adjuvant* to the antigen of interest. Another adjuvant used by immunologist is Incomplete Freund's Adjuvant (IFA), which is the oil additive without the Mycobacteria.

I have introduced CFA and IFA because the development of an experimental autoimmune disease in many animal models depends at least as much on the adjuvant as it does on the self-antigen. EAE (§150), for example, will most readily develop if the mouse or rat has been immunized with MBP emulsified in CFA; immunization to MBP emulsified in IFA, or to MBP in solution without any adjuvant, will not induce EAE. On the contrary, immunization to MBP in IFA or in solution can render the animal resistant to subsequent attempts to induce EAE by immunization to MBP in CFA. The mixture of the self-antigen in CFA is critical to the induction of other experimental autoimmune diseases too. The take-home lesson is clear: a self-antigen may activate an experimental autoimmune disease only within a particular context, CFA for example, and not IFA. A self-antigen administered in IFA or in solution seems actually to strengthen anti-autoimmune regulation (§153, §167). Now the only difference between CFA and IFA is the presence of dead Mycobacteria in the CFA. Thus the Mycobacteria in the CFA provide the critical context for the disease. In other words, an autoimmune disease can be triggered by an immune response to particular self-antigens in concert with an immune response to an infectious agent.

The self-antigens, like MBP, associated with experimental autoimmune diseases seem to be included within the set of homuncular self-antigens (§150). So we may pose the question: How does the immune response to the Mycobacteria in the adjuvant turn physiological autoimmunity into an autoimmune disease? It's a matter of interpretation, of immune decision-making. The Mycobacteria and the immune response to the Mycobacteria add *predicate* signals that turn the self-antigen into a compelling target for attack in the immune dialogue (§137). The string of signal molecules associated with the CFA, an ersatz infection, influences the nature of the autoimmune response made to the self-antigen. Disease results from the decision-making process; no more and no less. Other dead bacteria, cytokines, and true infections too can supply signals to modify immune decisions regarding homuncular self-antigens, and so turn health into disease. The pattern of signals within which the self-antigen is embedded can sway the response from a physiological to a pathological attractor.

In closing, let us note that infectious agents can even supply ersatz self-antigens along with predicate signals; molecules produced by infectious agents can mimic self-antigens of the host and present them in the context of the infection. CFA administered without a self-antigen can induce autoimmune arthritis in susceptible strains of rats; the Mycobacteria contain molecules that mimic the structure of molecules normally present in the joints of the rat. So the immune response to the Mycobacteria can activate autoimmunity to self-mimicking antigens in the context of the predicate signals of infection. In attempting to rid the body of the Mycobacteria, the immune system attacks the joints. Reactive arthritis and possibly other infection-associated autoimmune diseases in humans are thought by some to be triggered by self-antigen mimicry. Autoimmune disease emerges from a string of mixed signals; the self-antigen becomes confused with a persistent infection. As we know from child-rearing, contradictory demands on a cognitive system induce confused pathological behavior. Conflicting signals can drive brains crazy and immune systems mad. Clearly, ongoing anti-autoimmunity is needed to keep autoimmunity physiologically on track despite the contradictory signals imposed by self-mimicking infections.

§164 The Tissues and Autoimmune Disease

The immune system is in constant dialogue with the tissues (§137), exchanging patterns of molecular signals (§140). So the tissues themselves contribute to immune decisions, and thus to autoimmune diseases. The same self-antigen may be expressed, for example, in two different tissues, yet autoimmune T cells directed to that self-antigen might cause disease in only one of the tissues. For example, we have seen that pathological autoimmunity to an enzyme present both in the eye and in the lung can lead to eye disease in one individual and to lung disease in another. The nature of an autoimmune disease depends not only on the state of the immune system; the target tissue can co-operate with the lymphocytes, or resist their activities depending on the state of the tissue and the signals it sends to its potential attackers. Regional differences in immunity are critical, but are just beginning to receive the attention they deserve from immunologists.

Figure 33 summarizes the factors that influence the transition of physiological autoimmunity into autoimmune disease.

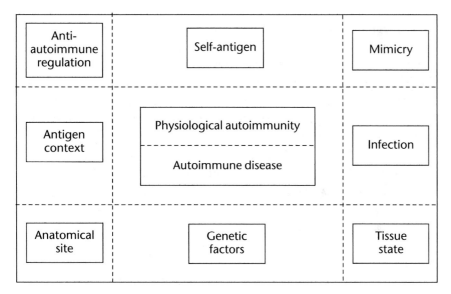

Figure 33: Factors that affect the transition between physiological autoimmunity and autoimmune disease

THERAPY OF AUTOIMMUNE DISEASE

Therapy of autoimmune diseases should be designed to reinstate healthy autoimmune regulation. Arrest of the disease process has been achieved by administering self-antigens in suitable contexts and by T-cell vaccination, the therapeutic vaccination of the patient against the antigen receptors of his or her autoimmune T-cells. Early treatment is essential to avoid irreparable damage to the target organ. Therapy can be tailored to the needs of the individual patient by monitoring the patient's response patterns.

§165 Autoimmune Therapy Defined

Defined goals distinguish fixing from meddling. What is the goal of autoimmune disease therapy? Consider that the patient suffers concurrently from two different disorders: the accumulated damage in the diseased tissue and the autoimmune process that produces the damage. At present, however, we only can hope to arrest the autoimmune process and stop the progression of the damage. Unfortunately, success in arresting the damaging process, of itself, will not repair the damage that has already been done. The restoration of tissue structure and function is a problem of another kind. Ideally then, the disease process should be stopped before damage becomes irreversible. What treatments have shown promise?

§166 Shifting Attractors

Think of treatment as control. One way to control a system, as we discussed, is to supply the system with information or energy to which the attractor interactions of the system are sensitive (§24). An effective control element will drive the system into a desirable set of interactions, into a healthy basin of attraction, as it were. The administration of either of two types of agents has been found, at least in experimental animals, to arrest autoimmune damage and restore physiological autoimmunity: self-antigens and the autoimmune receptors that recognize them.

167 Autoimmune Therapy by Self-antigens

Recall that the administration to experimental animals of a self-antigen within a context of adjuvant attack signals (§163) can transform
physiological autoimmunity into autoimmune disease. What information might turn the system the other way and transform it back into a
healthy set of autoimmune interactions, into a healthy attractor? Why,
the same self-antigen, but now framed in a healthy context of predicate signals. Re-educate the wayward lymphocytes. The good news is
that autoimmune lymphocytes are educable; various autoimmune
diseases in experimental animals may be arrested by administering self-
antigens. The bad news is that the treatments may need to be tailored
to individual patients; not all ways of administering self-antigens work
for all individuals. Let us start with the good news, and then discuss
individualized treatment.

Most encouraging is the fact that a disease involving a set of self-
antigens may be turned off by treating the animal with just one of the
antigens in the set. The diabetes in NOD mice, for example, is associated with heightened autoimmunity to many self-antigens including
insulin, a maintenance protein (hsp60), and an enzyme (glutamic acid
decarboxylase; GAD). Suitable administration of any one of these three
self-antigens to NOD mice can arrest the destruction of the insulin-
producing beta cells. Thus, the collective autoimmunity that
characterizes diabetes, and other diseases too, is sensitive to collective
control. In fact, treatment can be effective using a single peptide epitope of one antigen. Re-education of the autoimmune response to one
epitope of an antigen can apparently spread to other epitopes by by-
stander and other network connections (§123). My colleague Dana
Elias and I have developed a peptide treatment of autoimmune
diabetes, which is now being tried in human patients (see D. Elias and
I. R. Cohen, 1994).

Not only may native self-peptides effect therapy; peptides can be
altered by amino acid substitutions (altered ligands; §108) to enhance
their therapeutic efficiency. Indeed, Michael Sela, Ruth Arnon and
their colleagues in Israel have developed a random peptide mimic of
the self-antigen MBP(§150), which has been approved for use to treat
persons suffering from multiple sclerosis.

As you would expect, the context in which the antigens or peptides are administered is critical; the pattern of accessory signals endows the antigen with meaning (§127, §137). Hence, the therapeutic context must avoid adjuvant signals that might aggravate the autoimmune attack, and include, if possible, signals that predicate arrest of the attack. Some investigators are studying the therapeutic advantage of adding to the self-antigen T2-type cytokines that signal suppression of immune reactions (§120).

In addition to accessory signals, the antigen context includes the dose of antigen, the body site and the schedule of administration. The site of administration is important because, as we discussed, different anatomical sites manifest different immune requirements, and each site is programmed by its resident macrophages and lymphocytes to respond differently (§94). The gut, for example, is known to process ingested antigens in a way that leads to suppression of damaging immune responses. One may imagine that it is desirable *not* to mount a strong effector immune response to the food one eats. Howard Weiner and his colleagues in Boston have taken advantage of this property of the gut to design therapies for various autoimmune diseases based on *feeding* relevant self-antigens to persons with the disease. The myelin antigen MBP, for example, is fed to persons with multiple sclerosis, and insulin to persons developing diabetes.

We might borrow a metaphor from the nervous system to exemplify the idea of using a therapeutic context to modify autoimmune behavior. Recall the story I told of the crying baby and the smiling pediatrician (§68). The self-antigen is like a face that attracts the attention of autoimmune T cells; how the T cells respond to the face depends on the context, the pattern of predicate signals in which the face is seen. By administering the self-antigen in the right context, we show the autoimmune T cells the self-antigen face in the pattern of a smile. An angry self-antigen face activated the disease; a smiling self-antigen is medicinal. Or look at it this way: the self-antigen administered in an unequivocally benign context dispels the confusion of conflicting signals that triggered the disease.

Metaphors may enchant the mind, but metaphors are not science; we have yet to elucidate the molecular mechanisms by which self-antigens re-program the nature of the autoimmune response. Fortunately, the concept of physiological autoimmunity provides a rationale for thinking about the experiments we need to do to clarify the matter.

168 Activating Anti-Autoimmune Regulation: T-Cell Vaccination

The direct activation of anti-autoimmune regulation is another way to arrest the progression of an autoimmune disease. In fact, my own thinking about autoimmunity was decisively influenced by the unexpected discovery of regulation of autoimmune disease by a maneuver we have called *T-cell vaccination* (see A. Ben-Nun, H. Wekerle, I. R. Cohen, 1981, and the book edited by Jingwu Zhang and Jef Raus, *T Cell Vaccination and Autoimmune Disease*, 1995).

Avraham Ben-Nun, Hartmut Wekerle and I succeeded, in 1980, in isolating from rats with the autoimmune disease EAE (§150) pure cultures of T-cell clones responsive to MBP. Importantly, these cultured T cells were capable of producing EAE upon transfer into the bloodstream of otherwise healthy recipient rats. The activated T cells were able to make their way to the brain and spinal cord, causing inflammation and paralysis, even unto death. Thus we had in hand the T-cell agents of EAE – in fact, the original T cells are still in culture and can still cause EAE.

This T-cell culture technology, by the way, was used to prove that the immune systems of healthy individuals contained autoimmune T cells. The autoimmune T cells isolated from healthy rats, upon activation in culture, were shown to be capable of causing autoimmune disease in otherwise healthy recipients. The idea of the immunological homunculus was stimulated, among other observations, by this finding that essentially the same autoimmune T cells obtained from EAE rats could be recovered from healthy rats. (Recall that the CST had asserted that this experimental result was not possible; §147.)

It had been discovered a century earlier by Louis Pasteur and his colleagues that the causal agents of infectious diseases might be attenuated and used as protective vaccines (we now know that vaccination works by educating the immune system; §139). Would it be possible, we wondered, to vaccinate an animal against an autoimmune disease by educating its immune system with attenuated autoimmune T cells? Could the immune system, in other words, be vaccinated against its own excesses?

(We are taught that science proceeds by the formulation and testing of reasonable hypotheses; and that is what at least some scientists do at least some of the time. But I have the impression that many of the most revealing experiments in biology are done less through hypothesis and more through play. Cultures of functional T cells are fascinating toys, and not only reagents.)

To our delight, T-cell vaccination was found to protect and even cure animals of a variety of experimental autoimmune diseases: not only EAE, but thyroiditis, arthritis, diabetes, and others. Of course, treatment of each disease requires a T-cell vaccine composed of autoimmune T cells with antigen receptors involved in the particular disease; anti-MBP T cells can vaccinate against EAE but not against thyroid disease, and anti-thyroglobulin T cells work against thyroiditis but not against EAE, and so forth.

The investigation of T-cell vaccination has led to many findings, most of them unorthodox at the time of first observation. The findings, which led me to conclude that anti-autoimmune regulators are part of the homunculus (§153), can be summarized as follows:

- T-cell vaccination activates regulatory T cells, some of which recognize the antigen receptors of the autoimmune T cells used for vaccination. Other regulatory T cells recognize other molecules expressed by activated T cells. The regulatory T cells, like all T cells, recognize peptides associated with MHC molecules (P-MHC compound ligand; §115). Peptides derived from the autoimmune T-cell antigen receptor can be used as vaccines in place of whole T cells.
- The regulatory T cells do not kill the autoimmune T cells, but rather suppress their activities. The suppressed autoimmune T cells can be recovered from the protected animal, and these T cells, upon activation, can again produce the autoimmune disease in other animals. How this reversible suppression occurs is yet unknown.
- B cells that produce antibodies to T-cell antigen receptors are also

activated by T-cell vaccination. But we don't know if these anti-bodies have any regulatory function.

- The regulatory T cells exist even without T-cell vaccination; T-cell vaccination merely activates existing mechanisms of regulation. The natural regulatory T cells are probably responsible for controlling the autoimmune T cells that live quietly in our bodies.

- The induction of an experimental autoimmune disease can of itself activate the regulators. Such natural activation may spontaneously abort or limit the autoimmune disease. A progressive autoimmune disease appears to be associated with a decline in the activities of existing regulator cells.

- Treatment of an autoimmune disease by administration of a self-antigen (§167) can activate anti-autoimmune regulatory T cells. Conversely, T cell vaccination can induce a shift in the cytokines produced by the autoimmune T cells that recognize the self-antigen. In other words, therapy with self-antigens and T-cell vaccination may activate a common network of regulation. The network connections between T-cell vaccination and self-antigen therapy need elucidation.

- Exposure to infection can also activate anti-autoimmune regulatory T cells. This could explain how some infections prevent autoimmune disease (§162).

At the time of this writing, T-cell vaccination is being applied to the treatment of multiple sclerosis in several centers in the United States and Europe.

Figure 34 summarizes the two approaches to therapy.

Activation of immune regulation and reinstatement of *physiological autoimmunity* can be induced by the administration of *self-antigens*, taking into account the *dose*, *dose schedule*, anatomical *site* and *context*. Alternatively, enhanced regulation can be induced by vaccination with autoimmune *T-cells*, *T-cell receptors*, or T-cell *activation signals*.

169 A Problem of Therapeutic Individuality

Whether the self-antigen is fed or administered in other ways, the dose and the treatment schedule are critical. Too much antigen or too little

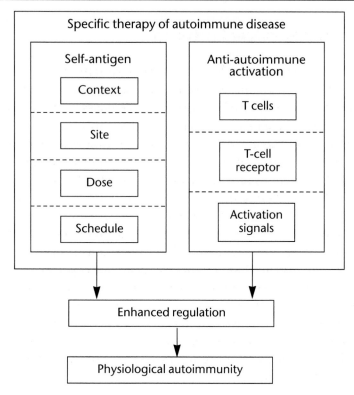

Figure 34: Specific therapy of autoimmune disease

antigen, given too often or not often enough, can elicit less than an optimal response, or might even aggravate the disease. Most vexing is the observation that different patients, even those with the same disease, might require different self-antigens, or different doses or dose schedules of a self-antigen to obtain the best treatment results. T-cell vaccination, too, is individualized treatment.

The need to tailor treatment to individual needs is understandable when we consider that, despite the large-scale uniformity of the immunological homunculus (§150), immune systems are individualized at the microscopic scale. Each person has self-organized his or her immune system in the light of a unique immune history; thus the state of the immune system is not likely to be the same in all patients suffering from the same autoimmune disease. At the microscopic scale, each patient can express a private pattern of disease, and private patterns need private treatments. Here as elsewhere, individuality can be problematic.

170 Measuring Individual Patterns

An autoimmune disease reflects the sorry state of the patient's immune system. The fundamental problem is a sick immune system; the disease, in effect, is a by-product. Put the immune system into a healthy response pattern, and the disease, of itself, will go away. Specific therapy amounts to treating the patient's immune system, or more precisely, the state of the system. A rule of good medical practice links effective treatment to correct diagnosis; you should first be aware of what needs fixing before you try and fix it. To cure an autoimmune disease safely and efficiently, it would be wise to know what's wrong with the system. Now, the state of a system is defined by the collective pattern of the states of the individual elements constituting the system (§17). To diagnose the diseased state of the immune system and to monitor the effects of our therapies, we shall have to develop the means to record and analyze the patterns of activity of collectives of immune agents: cytokines, antibodies, T cell populations, genes, whatever. By monitoring patterns, we shall be able to evaluate and adjust treatment as the disease evolves. Monitoring the response to treatment is the key to individualized therapy.

The call to analyze global patterns departs from the traditional interest of immunology in discrete immune responses. This approach also deviates from the traditional desire of drug companies and government agencies to have a single therapeutic agent bottled for standard dose administration to the whole patient population. Individualized therapy, and its attendant complexity, is looked upon with suspicion, if not horror.

Alas, the immune system is a complex system and its behavior emerges from co-operative interactions and patterns of signals. To ignore this complexity is to guarantee the continued failure of specific immune therapy; in any population of patients there are likely to be a sufficient number of non-responders to threaten the statistical significance of any standard clinical trial. Fortunately, Coutinho and his colleagues have initiated research into global antibody patterns (§151). The newly developed 'chip' technologies and the emerging field of bio-informatics promise to help immunology (and industry/government) measure the dynamic responses of individuals to specific treatments. We have to delineate new research goals and proceed to implement them. Might it be possible one day to prevent the eruption of an incipient autoimmune disease by the early treatment of an abnormal immune system pattern?

Figure 35 summarizes the use of global patterns to individualize treatment of autoimmune disease.

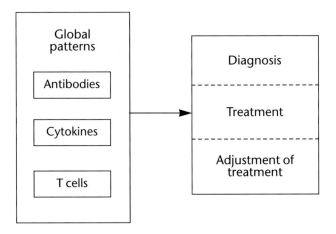

Figure 35: Global patterns in the diagnosis and treatment of individual subjects

AUTOIMMUNITY IS OF THE ESSENCE

Autoimmunity is not an aberration, as taught by the classical CST, but is at the heart of the immune system.

§171 Essential Autoimmunity

We can now sum up the situation. Far from being an accidental blowout or a necessary evil, autoimmunity is essential to the healthy behavior of the immune system: the ability to recognize foreign antigens is a product of primary selection for degenerately autoimmune T-cell clones; homuncular autoimmunity is organized within the system, and serves to enhance co-respondence, facilitate immune maintenance, fight infections, reject foreign tissues, and possibly destroy tumor cells.

Autoimmune diseases are among the costs we pay for a potent homunculus. The reinstatement of physiological autoimmunity provides a natural way to make the system right. Be that as it may, the immune system is organized by autoimmunity; autoimmunity is essential.

COMPARATIVE SUMMARY OF AUTOIMMUNITY

The CST and cognitive paradigms of immunity differ radically. What is a paradigm? Should the CST be replaced?

172 Paradigms

The cognitive view of autoimmunity, like the cognitive view of the immune system generally, is distinctly different from the views of the classical CST paradigm. How can we know which paradigm is preferable? Does it matter? What use is a paradigm anyway?

A scientific paradigm is a conception of reality adopted by a community of scientists that allows the community to proceed with a common program of research. Scientists who share a professional world-view of their joint subject matter can agree in large part on the questions worth investigating, on the technologies suitable to the experimental quest, on the interpretation of experimental results and natural observations, on the criteria for judging success, and on the dispensation of merit and honor. A scientific paradigm is a way of carving reality into segments suitable to the scientific enterprise (§57).

(Paradigms, of course, are not exclusive to scientists; every group needs a paradigm to organize its membership: religions, nationalities, professions, social and political movements, revolutions, and families. Individuals, too, acquire a feeling for the nature of the world and their place in it – which is a personal paradigm.)

Note that a paradigm is not merely a theory; a paradigm can hold divergent theories within its embrace. Science progresses, according to the philosopher of science Karl Popper, by falsifying its theories; a good scientific theory is one that can be disproved by the emergence of a contradictory empirical fact (see K. R. Popper, *The Logic of Scientific Discovery*, 1959). All agree, officially, that science closes in on the truth by discarding and replacing theories that no longer can account for the facts. Paradigms, although more robust than theories, must eventually become outdated and are discarded when the facts discovered through scientific inquiry can no longer be subsumed under the paradigm without insulting the collective intelligence.

In practice, however, a scientific paradigm often continues to thrive in the face of discoveries that seem to contradict its central tenets. Scientists honor Popper's philosophy, but don't really like to see honored ideas falsified. (Our instinctive wish to preserve ideas is, after all, a guardian of human culture; §79.) The philosopher of science Thomas Kuhn, in contrast to Popper, teaches that a paradigm is held dear to the hearts of the community of its adherents for reasons that are largely psychological; paradigms give meaning to life and time. Tribes are held together by their mythologies, scientific disciplines by their paradigms (see Thomas S. Kuhn, *The Structure of Scientific Revolutions*, 1970).

Kuhn quotes the physicist Max Planck who sadly reported that a new scientific truth finally replaces an out-of-date paradigm, not through its compelling logic, but only when the last of the adherents of the old paradigm have retired from the field, and a new generation, freer of the outmoded tradition, has gained ascendancy. Should the CST worldview be replaced? Perhaps the CST and the cognitive theory are merely contending interpretations of a common paradigm?

§173 Paradigm Shifts

Had a divergent interpretation of autoimmunity been the only difference between the cognitive theory outlined here and the classical CST, I would guess that both theories could continue to live side-by-side within the same paradigm. After all, whether autoimmunity is deleted or regulated is, at least to an outsider, a parochial point of disputation, not a world-shattering crisis. Indeed, the point has not bothered the community of immunologists who, irrespective of deletion or regulation, can proceed to ferret out the clockwork of the immune system gene by gene and molecule by molecule. Two other issues, however, will, I think, be decisive.

The first issue is theoretical. The CST revolves around one central attractor, that of receptor specificity; the CST begins with specificity as its center of gravity. Specificity, it believes, is the natural endowment of the antigen receptors. To my mind, therefore, the CST is endangered by the realization that antigen receptors are intrinsically degenerate. Specificity is not a given, but an outcome. Hence, an immune paradigm that cannot account for the system's down-stream construction of specificity must eventually be doomed to rejection for inadequacy. The cognitive paradigm is more fitting because it proposes a theory of specificity that highlights the problem of degeneracy and marks out a new territory for research. The cognitive paradigm, like the CST, will have done its job well if it, too, succeeds one day in driving itself out-of-date.

The second issue that could sink the CST is practical. Immunology is essentially a clinical science; it prescribes a way to understand and deal with health and disease. Now the CST recommends the destruction of the autoimmune clone as the way to cure an autoimmune disease. The cognitive paradigm, in contrast, suggests the importance of monitoring immune system patterns and prescribes the positive activation of autoimmune regulation. The paradigm that leads to the cure of auto-immune disease will prevail.

Chapter 6
On Tending Adam's Garden

Chapter 6
On Tending Adam's Garden

ADAM

Adam is the first individual. His history is an insight into the self.

§174 The Name of the Book

We have reached the end of the book, and you might be wondering why I chose the title *Tending Adam's Garden*. What does a cognitive paradigm of the immune system have to do with Adam? True, I invoked Adam and Eve to add body to a concept of knowledge (§43, §90), and I referred indirectly to Adam when we discussed the naming of names (§5, §149). But merely referring to Adam is hardly sufficient reason to give him top billing. I placed Adam and his Garden in the title because the story of Adam suits our subject; cognition is about individuality, choices, consequences, and the importance of tending. Adam is each of us, his Garden is our world, and Adam's story, like it or not, is our story. To your own musings, let me add some additional material about Adam drawn from the Jewish literary tradition – which, after all, is a first-hand account.

§175 Individual Adam

According to the Talmud, Adam embodies the idea of individuality. The Talmud is a compendium of Rabbinical thought that has its beginnings some time after the canonization of Jewish Scripture well over 2,000 years ago and extends for some 800 years thereafter; from about 300 BC to about AD 500 (see The Talmud: *The Steinsaltz Edition/Commentary by Adin Steinsaltz*, 1989). The Talmud aims at establishing the principles, practice and spirit of a way of life. Rather than enunciating abstract principles, the Talmud makes its points by recounting case histories and concrete portraits, often communicated as interpretations of Scripture.

In the section dealing with the adjudication of capital crimes, the Talmud inserts into the proceedings of the court the idea that human life has cosmic significance. This is done to impress the witnesses with the gravity of their testimony (*Tractate Sanhedrin*, 37b). The creation of Adam is the hammer that drives home the point. Why, asks the court, did the Creator begin with Adam alone; if He or She wanted a world full of people, why bother starting with only one? Adam was created as a singularity, proclaims the court, to teach his descendants that each person is the equivalent of a unique creation; whoever saves a human life rescues a world and whoever causes the loss of a human life destroys a world. Every person, including the accused, is the indivisible center of a world. Thus, continues the Talmud, every person must acknowledge daily that the world was created for him, personally (and act commensurate with that awful responsibility, adds Rashi, the noted commentator of the eleventh century; apparently the Rabbis believed self-respect is a better goad than is a sense of sin).

Our analysis of cognition adds biologic support to the Talmud's hermeneutics; each individual fashions a unique world out of his or her unique somatic experience. Therefore, no individual is redundant, ever. The Talmud, we might assume, knew about the cognitive brain; we have added the cognitive immune system to the armor of individuality.

The obverse side of individuality, of course, is diversity. Human diversity, says the Talmud in the same context, attests to the glory of the Creator: out of a single mold, Caesar mints coins for the world, and all the coins are the same; the Creator has minted a world of humans from a single Adam (even Eve, we are told, was cloned from Adam – with a sex chromosome modification, we may presume), yet no two humans are identical. The tale of Adam, like the message of this book, is a tale of cognitive individuality. One's views about the identity of the Creator or the role of evolution cannot gainsay the truth of that message.

§176 Self-Organizing Adam

Let us return to the biography of Adam. The continuing evolution of Adam emerges through mutual interactions with Eve, the serpent, and the rest. The cognitive self-organization of Adam and of Eve is marked by decisions and by internal images; recall the tree in the center of the garden, the awakened sense of nakedness, the hiding, the blame and the recriminations. The self-organization of the couple and their children constitutes a strategy, not always successful, for dealing with the world. Note that Adam's tale personifies history, the irreversibility of time; eviction from the Garden and the death of Abel are permanent – there is no way back from somatic responsibility, no cyclical starting over (§37).

§177 The Self

The story of Adam is the story of the self. What is the self? The self cannot be a fixed material substance; all of our molecules, like those of a river, are in constant flux (§12). There is no absolute antigenic distinction between the self and the not-self; the immune system can recognize and respond to either (§149). The DNA code is not the self; one's genes, by themselves, have no expression and no meaning (§41). The mind is not a stable entity; the flux of ideas from moment to moment expresses the inconstancy of the brain. Perhaps the soul serves as the seat of the self; except nobody has ever measured, or even seen one. I would like to argue that the self cannot be reduced to a core reality; the self, like life, is an emergent property (§12).

The self is composed of interactions, self-organization, history, and memory. The self, like the story of Adam, is a story of interactions with a world of people and things; the self emerges as a string of self-organizing events. One's body, though its composition may change, is the seat of the action – the context of the body is like the covers of the book within which the story unfolds.

The self is a coherent entity by virtue of the history of interactions it records. History is the seat of the self; the progression of an irreversible history centered in one body is the self. Irreversibility is the mark of self-organization; the interactions have a direction; the interactions self-organize – the organized composite is the self.

The crown of the self is memory. Memory is the trace left by a life – the river bed that records the flow of the river (§28), or perhaps the wake in the sea made by the passing boat (§9). Memory is the impact of life on one's own brain and on one's world. Life persists in the memories it has created. Yizkor, the Jewish prayer for the dead, is a prayer for memory: may the Eternal remember. The way you live is the world you build, and the world you build is the memory of your life (§42). Just like Adam. Adam is the memory of Adam. The self is a-story-in-a-context. Adam, of course, is only a metaphor; you are free to entertain your own interpretation.

§178   The Tree of Knowledge

Central to the story of Adam is the tree of knowledge. The knowledge that comes of eating its fruit marks a decisive change in the history of Adam and his offspring. We have discussed knowledge as a form of cognitive doing (§43, §90). Can we gain any further insights by considering the type of tree that furnished the fateful fruit? Scripture says nothing about the species of the tree. For reasons unknown to me, the tree of knowledge, at least in the West, is believed to have been an apple tree. But the Talmud makes three suggestions (none of them apple). One sage proposes that the tree of knowledge was a fig tree, the second that it was a grapevine, and the third that it was wheat (*Tractate Sanhedrin*, 70a). These, to my mind, are more interesting than the apple.

The fig signifies sexuality, which always interests humans; and that's the point. Human sexuality, with some early intimations among the higher primates, is uniquely divorced from procreation, and empowers human motivation in a way unique to our species (see S. Freud, *Sigmund Freud Collected Papers*, 1959). The Rabbis appear to pre-empt Freud in telling us that the connection between sexuality and knowledge is not trivial. In any case, human sexuality is an attractor that, for good and bad, has separated us from the rest of biologic evolution.

The grapevine points to wine and its power of intoxication. What are the Rabbis telling us here; is intoxication knowledge or is it anti-knowledge? Be that as it may, wine-making is an early venture in bio-technology. Is bio-technology usurpation? Is it unnatural knowledge?

Wheat is most momentous in human affairs. The domestication of wheat and the ensuing agricultural revolution detached us from our natural condition as a species of hunters and gatherers, changed the structure of the human population, and led to a conquest of nature that now threatens to transform our world irrevocably.

In short, the Talmud views the incident of the tree of knowledge as a metaphor for the attributes (impious?) of humankind that have set us apart from all other creatures – the powers of sexuality and of cultural development. Knowledge, nevertheless is not bad necessarily. Rashi explains that, beyond the general blessing of life given to all living creatures, the added blessing bestowed on Adam at his creation was 'the accruement of conscious knowledge and speech' (see Rashi on Genesis 2:7). Adam, therefore, cannot be said to have seized the power of cognition through trespass. (The later misuse of speech did bring the children of Adam to the tower of Babel, which is another story.) But blessing or not, Adam's brand of knowledge has changed the course of the world. Which brings us to our last topic: how are we to manage what's left of the Garden? We, like our immune system, have to worry about maintenance.

TENDING THE GARDEN

The world is no longer the garden we evolved to live in. Maintenance means tending.

§179 In the Beginning

Modern humans have been around for about 100,000 years. The evolutionary line that culminated in humans, however, branched off from the other primates perhaps 5–10 million years earlier. But immunologically, the precise time humans have been evolving makes little difference. Like the other higher primates, we have spent almost all of our evolutionary time foraging about in small groups of some tens to a few hundred individuals in an ongoing search for food. We don't know how many people populated the world during those formative years, but the density of the human population was certainly very sparse. And most of the people were probably young; by the evidence of skeletal remains, few if any managed to survive beyond the age of 40. The agents of infectious disease which made their living on our persons were adapted to surviving within a sparse population of wanderers. The chronic diseases of old age were no problem, because there were no old people. This was the world within which our immune systems evolved.

§180 Harmless Parasites

Defending the body against pathogenic infectious agents is one of the major tasks of the immune system. Reaching an accommodation with harmless infectious agents, or at least with some of them, is no less important. In fact, harmless parasites are harmless *because* of the accommodation they have reached with us. Since we normally accommodate such parasites, we can call them *normal* parasites.

Our skins are covered with normal bacteria, many of our cells harbor normal viruses, and our large bowel is like a fermentation chamber housing billions and billions of normal bacteria. Normal infectious agents, for the most part, live with us in these situations in peace. But in other situations, many of our normal parasites manifest the capacity to harm and kill. Normal skin bacteria such as Staphylococci and Streptococci can penetrate the skin to cause boils, and enter the blood to cause death. Cytomegaloviruses (CMV) and other normally harmless viruses, which most of us carry in our bodies for a lifetime, can erupt in the elderly or debilitated and kill the host. The normal bacteria of the gut can invade the blood from a strangulated (no blood supplied) bowel, or from a surgically manipulated urinary tract and cause lethal shock. How do these agents cause disease, what keeps them harmless, and what controls the transition to harm? The host, the parasite, the immune system and the laws of evolution do.

§181 Host–Parasite Co-Evolution and Complexity

The key to host–parasite accommodation is co-evolution. Remember the first two laws of evolution: available energy and space will be occupied (§31). The emergence of any creature, us included, provides energy and space for exploitation by other creatures; the more complex the creature, the more provisions there are for other creatures to exploit. The more types of cells and tissues there are, the more ecological niches there are. So the laws of evolution ensure that creatures will evolve to exploit the energy and the space that is supplied by our complexity; such an exploiter will be defined here as a parasite.

It is probably no accident that creatures with complex tissues also have complex immune systems. Lymphocytes, along with their ability to recognize antigens and their cognitive self-organization, first appear in creatures constructed of many differentiated tissue types – the vertebrates. The plants and the invertebrates, with their few and relatively simple tissue types, have populated the earth and thrive to this day without the benefit of even a single lymphocyte. An innate, germ–line immune system of macrophage-like cells satisfies them completely. These lymphocyte-less creatures also manage without a closed blood circulatory system. Apparently, the evolution of creatures with complex tissues required the evolution of a closed blood system and an adaptive immune system. Maintenance and protection become complicated.

Here is a note of immunological humility; it's not that the cognitive immune system of the vertebrates is more effective or more improved than are the simple innate immune systems of the plants and the bugs. It's just that the complex composition of vertebrates requires extreme maintenance. Complexity dazzles, but costs; simple tissues and innate immune systems don't get cancer. The DNA code, as we discussed, has not necessarily 'improved' through natural selection, it has merely become more complex (§41).

How might a working accommodation between host and parasite evolve? Recall that a creature survives when it participates in a stable arrangement, an attractor (§30). The parasite and the host, therefore, have a common interest in stability. The parasite needs the survival of the host, at least for as long as it takes to make more parasites and infect the next host. (Below, we shall discuss what happens to the arrangement when the parasite does not need the survival of the host; §183 – If this possibility worries you, it should.). Teleologically, the host, for its part, enjoys a well-adapted parasite; it makes more sense to accommodate a relatively harmless parasite than it does to eject it and have the familiar parasite replaced by a potentially more dangerous stranger. An effective host strategy is to keep its niches occupied with familiar parasites.

Hence, a host and its well-adapted parasites tend to co-evolve to accommodate each other's needs. The host supplies food and shelter, and the parasite supplies an occupied niche, an immunological education (§162) and, in the gut at least, important vitamins. The host and the parasite are careful not to harm one another unnecessarily; the parasite does not invade vital tissues and the host does not eradicate the parasite, as long as it remains harmlessly in its assigned niche.

The host–parasite arrangement is maintained by a complex array of mutual signals; the parasite makes molecules for which the host has germ-line receptors (§114); such signals induce a strong immune response to any errant parasite that chances to leave its assigned niche and attempts to invade the body. The host, for its part, supplies molecules and structures needed by the parasite to occupy its assigned place.

Note that symptoms too are signals. Fever, muscle pains, lethargy, loss of appetite, sensitivity to light and the other miseries of illness are induced by immune molecules, and by parasite molecules too; these signals inform the central nervous system that the person needs rest and isolation. The fever also helps the body combat many infectious agents. The host–parasite relationship is well orchestrated.

§182 Mercy Killing

But the arrangement with normal parasites is not always harmless; we did mention that well-adapted bacteria and viruses can kill us, under certain circumstances. How do these parasites kill, and what are the circumstances?

Strange as it may seem, our normally harmless parasites can kill us by activating the immune system. The so called toxic shock syndrome can occur when otherwise harmless bacteria invade the bloodstream and stimulate immune and tissue cells to secrete massive amounts of cytokines (TNFα, IL-1, IFNγ) and other apoptosis-inducing molecules (§97). The immune molecules kill the host that has secreted them. Indeed, the very same immune system molecules that kill the host are the agents that function to destroy invading infectious agents. Life and death turn out to be a matter of timing and quantity; a lesser amount of cytokines produced earlier in the invasion will destroy the infection and cure the host; a greater amount of cytokines late in the invasion will kill the host (and the host-trapped parasite too). It's as if the immune system and the parasite complied in terminating their arrangement. The parasite can survive if it has managed to reach a new host either before or shortly after the death of the host. The host, unlike the parasite, is an individual, and is lost for ever.

Under which circumstances do our normal parasites kill us? Usually when we are quite sick with some other illness. Gut bacteria invade through dead bowel tissue. Latent viruses emerge from body cells when the immune system is poisoned by drugs or weakened by old age. The very young, in addition to the old and sick, are choice targets of normal parasites, particularly when the young are malnourished or debilitated. Does the host–parasite arrangement include mercy killing?

Perhaps the idea is not so bizarre: a death arrangement, if such exists, between the host and its well-adapted parasites calls to mind an aspect of the relationship between a predator and its prey. Zebras and lions, for example, organize the hunt using an array of signals understood very well by both sides. Zebras and lions pretty much ignore each other until the lions announce their readiness to hunt with roars and body posture. The zebra herd responds with a series of displays that easily show the lions which members of the herd are infirm or unfit. The lions pursue only the marked zebras and so gain a meal free of the cost in energy and potential broken bones involved in trying to capture and kill a fit zebra. The zebras, for their part of the bargain, benefit from having the herd culled of sick and weak members which might endanger the healthy members while competing for scarce food and weakening them all. Old zebras are not sent out to die on the ice; the lions do the job. A fascinating description of predator–prey signaling and complicity may be seen in *The Handicap Principle: A Missing Piece of Darwin's Puzzle*, by Amotz and Avishag Zahavi, 1997.

Could it be that immune system suicide is programmed to cull from the group unfit individuals who might threaten group survival? An immune system no longer well enough to maintain a healthy individual kills its unfit self, using the normal parasites as an indicator; the time has come to commit suicide when normally harmless bacteria can invade the bloodstream. The species, like a herd of zebras, keeps itself fit by calling in the predators to clean up, and helping them do it too. Even parasites tend their Gardens.

Evolution is supposed to work on individual survival, though, not on group survival (§79). How could a species have ever evolved an immune program to kill the individual? But the controversial issue of altruism, the individual's self-sacrifice for the good of the group, is beyond our present scope. Just note that the immune system is a contractor of apoptosis for sick cells (§97), and for sick individuals, too, when the need arises.

§183   The Danger of Independent Parasites

Now we can define the conditions in which it pays for a parasite to be harmless. A parasite will tend to be harmless if its way of life is such that the parasite needs a healthy host to keep itself alive. Note the *if*; whether or not the parasite needs a healthy host is critical to our understanding of the evolving relationship between a host and its parasites. A parasite that needs a healthy host for its own survival will tend to be harmless and keep its host healthy; a parasite that does not need a healthy host may be quite harmful. And a parasite that has evolved to exploit a sick or dying host for its propagation will surely be harmful. For a full treatment of the subject, you can read *Evolution of Infectious Disease*, by Paul W. Ewald, 1994. Here, I shall only bring a few examples to illustrate the idea that some parasites are really out to kill us, provided the conditions reward it for doing so. Adaptation is *fittedness*; what works, works (§33). Obviously, it is in our interest to reward parasites for moderation, not for virulence (from *virus, poison* in Latin). Virulence tends to emerge when the human host is accidental or optional for the parasite, or when the parasite is transmitted through a *vector* (a carrier, from the Latin *vehere*, 'to carry').

Accidental host. Rabies is an apt example of a dead-end infection. The rabies virus normally survives by infecting bats and other wild creatures; the infection of a person by a rabid dog is a regrettable accident in which the virus has no stake. Bacteria that make their living in the soil, such as anthrax or tetanus, can also kill you by mistake. There is no co-evolution here, and, hence, no genetic moderation of the agent or resistance of the human; the parasite really lives somewhere else.

Optional host. Infectious agents that make their living by infecting other animals, but which have the option of infecting humans can also kill people with impunity. An example is plague. Although infected humans can disseminate plague, the plague bacterium survives by living a more moderate life in its non-human hosts.

Vector-borne parasites. Parasites that get from one person to another by way of carrier agents can benefit from making the human host sick, if a sick host can assist the process of parasite dissemination. Vector-borne parasites are fairly common.

- *Insect vectors*: Malaria, for example, is spread by mosquitoes, not by mobile people. In fact, a sick, immobile person whose blood is teeming with malaria organisms is an ideal meal for mosquitoes, who can pass on the infection. Insect-borne parasites of many types benefit from virulence.
- *Water vectors*. Cholera, too, does not need mobile hosts to survive; a contaminated water supply does the job. And explosive diarrhea keeps the water contaminated – provided the sewer system has access to the drinking water. Thus a bad water system selects for virulence. By the same token, clean water selects for parasite moderation because the parasite will then need healthy, mobile hosts for dissemination.
- *Hospital vectors*. Hospitals are characterized by concentrations of sick patients, virulent infectious agents, contaminated equipment, and attendants who go from patient to patient. The sicker the patient, the more the patient is surrounded by attendants. You can see how dissemination by equipment and attendants rewards virulence.
- *Population density*. A high density of people is a vector that rewards virulent parasites. A sick person has no difficulty spreading an infectious agent when the sick person is in a crowd. Indeed, the sicker the person, the more infectious he or she is likely to be at close quarters. Historically, the agents of measles and influenza, which are transmitted by acutely sick patients, probably could not have taken root in humans before the agricultural revolution and the resulting baby boom supplied the viruses with large numbers of freshly susceptible people.

§184 East of Eden

It might have occurred to the reader that most of the conditions favoring virulence could not have operated when Adam was still residing in Eden, in the context of a small isolated family. The agricultural revolution, the fruit of knowledge, led to an enormous growth of a settled human population, intense contact with domesticated animal vectors, poor sanitation, contaminated and stagnant water, crowded cities, hospitals, overuse of antibiotics, armies, wars, air travel, unsafe sexual promiscuity, old age, prolonged debility, malnutrition, overweight, physical unfitness, chronic tension, and all the other plagues of the present world that foster virulent infection and disease.

All in all, life in Eden was healthy, though short. The remains of humans who died before the agricultural revolution seem to be quite free of the stigmata of chronic disease (see *Digging for Pathogens: Ancient Emerging Diseases – Their Evolutionary, Anthropological and Archaeological Context*, edited by Charles L. Greenblatt, 1998). People in Eden apparently died quickly – of starvation when they could no longer gather food, of injury on the hunt, or by the teeth of cave bears, when hunted. Human bones begin to show chronic disease, along with old age, after the emergence of agriculture, sedentary living, and population growth. The continuing growth of the population, accelerating destruction of the natural environment, global pollution and global warming are not very encouraging. Wheat, the fruit of the vine, and the fig have brought us into a strange Garden. Our brains have to solve problems not previously encountered during the millennia of our biologic evolution. The ever-increasing complexity of human culture, along with its rewards and punishments, has created new opportunities for parasites and, consequently, new challenges for the immune system (§32).

§185 Tending

Even the best of gardens need tending. Adam was charged with caring for his Garden, even before he got into trouble (see Genesis 2:14). How can we stay out of worse trouble? We have discussed information, meaning, energy, order, self-organization, evolution, decision-making, goals and images; peoples and their governments, we hope, will use these factors and tend the Garden, such as it is, with wisdom. The immune system certainly can be reinforced by clean air, clean water, effective sewerage, and the rational use of antibiotics; such measures will tend to encourage the evolution of moderate parasites rather than virulent parasites. But to immunize against cancer, design successful organ transplantation, prevent and cure autoimmune disease, and vaccinate against emerging infections, it may be helpful to deal with the system in its own cognitive terms. Let us learn to diagnose the molecular patterns of the immune system (§170) and influence immune behavior using the molecular signals of the system's own chemical language (§137). If we could administer the right string of immune signals, we might be able to turn on or off the immune response as we see fit. Two cognitive systems are better than one.

References

Atlan, H. and Cohen, I. R. (1998) 'Immune Information, Self-organization and Meaning', *International Immunology* 10, pp. 711–17.

Ben-Nun, A., Wekerle, H. and Cohen, I. R. (1981) 'Vaccination Against Autoimmune Encephalomyelitis with T Lymphocyte Line Cells Reactive Against Myelin Basic Protein', *Nature* 292, pp. 60–1.

Boehm, U., Klamp, T., Groot, M. and Howard, J. C. (1997) 'Cellular Responses to Interferon-γ', *Annual Review of Immunology* 15, pp. 749–95.

Burnet, M. (1969) *Self and Not-Self*, London: Cambridge University Press, p. 230.

Cohen, I. R. (1992) 'The Cognitive Paradigm and the Immunological Homunculus', *Immunology Today* 13, pp. 490–4.

Cohen, J. and Stewart, J. (1994) *The Collapse of Chaos*, New York: Penguin Books.

Dawkins, R. (1976) *The Selfish Gene*, Oxford: Oxford University Press.

Edelman, G. M. (1987) *Neural Darwinism*, New York: Basic Books.

Elias, D. and Cohen, I. R. (1994) 'Peptide Therapy for Diabetes in NOD Mice', *Lancet* 343, pp. 704–6.

Ewald, P. W. (1994) *Evolution of Infectious Disease*, New York: Oxford University Press.

Freud, S. (1959) *Sigmund Freud Collected Papers*, New York: Basic Books.

Gould, S. J. (1977) *Ever Since Darwin. Reflections in Natural History*, New York: W. W. Norton, p. 36.

Greenblatt, C. L. (ed.) (1998) *Digging for Pathogens: Ancient Emerging Diseases – Their Evolutionary, Anthropological and Archaeological Context*, Rehovot, Israel: Balaban Publishers.

Harel, D. (1987) 'Statecharts: A Visual Formalism for Complex Systems', *Science of Computer Programming* 8, pp. 231–74.

Kuhn, T. S. (1970) *The Structure of Scientific Revolutions* (2nd edition, enlarged), Chicago: The University of Chicago Press.

Manna, Z. and Pnueli, A. (1992) *The Temporal Logic of Reactive and Concurrent Systems*, Berlin: Springer-Verlag.

Moalem, G., Leibowitz-Amit, R., Yoles, E., Mor, F., Cohen, I. R. and Schwartz, M. (1999) 'Autoimmune T Cells Protect Neurons from Secondary Degeneration After Central Nervous System Axotomy', *Nature Medicine* 5, pp. 49–55.

Myer, E. (1961) 'Cause and Effect in Biology', *Science* 134, pp. 1501–6.

Pinker, S. (1994) *The Language Instinct*, New York: William Morrow and Company.

Plotkin, H. (1995) *Darwin Machines and the Nature of Knowledge*, Cambridge, MA: Harvard University Press.

Podolsky, S. H. and Tauber, A. I. (1997) *The Generation of Diversity. Clonal Selection Theory and the Rise of Molecular Immunology*, Cambridge, MA: Harvard University Press.

Popper, K. R. (1959) *The Logic of Scientific Discovery*, New York: Basic Books. Republished 1992 London: Routledge.

Rook, G. A. W. and Stanford, J. L. (1998) 'Give Us This Day Our Daily Germs', *Immunology Today* 19, pp. 113–16.

Steinsaltz, A. (1989) *The Talmud: The Steinsaltz Edition/ Commentary*, New York: Random House.

Strohman, R. C. (1997) 'Epigenesis and Complexity. The Coming Kuhnian Revolution in Biology', *Nature Biotechnology* 15, pp. 194–200.

Tauber, A. I. and Chernyak, L. (1991) *Metchnikoff and the Origins of Immunology. From Metaphor to Theory*, New York: Oxford University Press.

Zahavi, A. and A. (1997) *The Handicap Principle: A Missing Piece of Darwin's Puzzle*, New York: Oxford University Press.

Zhang, J. and Raus, J. (eds) (1995) *T Cell Vaccination and Autoimmune Disease*, Heidelberg: Springer-Verlag.

Index